THE RICARDO STORY

The Autobiography of Sir Harry Ricardo, Pioneer of Engine Research

Second Edition

SAE Historical Series

Published by:

Society of Automotive Engineers, Inc.

400 Commonwealth Drive

Warrendale, PA 15096-0001

First published as *Memories and Machines: The Pattern of My Life* by
Constable & Co. Ltd. in 1968. Printed in Great Britain.

Reprinted in 1990 by Ricardo Consulting Engineers Ltd.

Library of Congress Cataloging-in-Publication Data

Ricardo, Harry R. (Harry Ralph), Sir, b. 1885.
   [Memories and machines]
   The Ricardo story : the autobiography of Sir Harry
Ricardo, pioneer of engine research.—2nd ed.
      p.   cm.—(SAE historical series)
   Includes index.
   ISBN 1-56091-211-1
   1. Ricardo, Harry R. (Harry Ralph), Sir, b. 1885.
   2. Mechanical engineers—Great Britain—Biography.
   I. Title. II. Series.
   TJ140.R5A3 1992
   621'.092—dc20
   [B]   91-43598
                                       CIP

ISBN 1-56091-211-1

Contents

	Preface to Second Edition	*page* 7
	Foreword	9
1	Early Life	13
	Activities in London – my family and friends – holidays and growing up	
2	Family Background	28
3	Schooldays	38
4	Mechanical Engineering at the Turn of the Century	51
	Steam engines – railways – gas engines – Hornsby Ackroyd and Diesel – hot air engines – the horseless carriage	
5	Some Early Endeavours	69
	Steam engines and my first internal combustion engine	
6	Cambridge	81
	The fuel economy run – research work with Hopkinson	
7	The Dolphin Venture	96
	Design and manufacture of an engine and some motor cars	
8	Start of a Career with my Grandfather's Firm	113
	Mechanical engineering in a civil engineering context – early work on the internal combustion engine	
9	The End of the Beginning	130
	State of the art in 1914 – ships – trains – diesel engines – motor cars – aircraft	
10	The First World War	147
11	The First Tanks	165
	Controversy at their birth – their engines	

12 Formation of my Company   *page* 182
    Working with Hopkinson again – the Air Ministry

13 A New Direction   198
    Post-war – transformation of the I.C. engine –
    development of the Turbulent Cylinder Head – petrol
    fuel investigation – the first Atlantic crossing – alcohol
    for engine fuel

14 Some Episodes from the New Company   213
    Bearings and lubricants – a strange accident that
    confirmed a theory – sleeve-valve engines – the airship
    controversy – engines for the R100 – our early clients

15 Patents and Royalties   231
    A patent action that meant life or death to my hopes

16 Some Frivolous Pursuits   243

    Postscript by Cecil French   259

    Appendix A – Biographical Notes   267

    Index   277

Illustrations

Harry R. Ricardo   *frontispiece*

Harry Ricardo, aged about six   *facing page* 12

Harry Ricardo, in his early twenties (a) greasing
rear wheel bearings, (b) enjoying his pipe
aboard motor launch on upper Thames   13

Sir Dugald Clerk   112

Professor Bertram Hopkinson, from the
portrait by Arthur T. Nowell, painted in 1911   112

Dr Frederick Lanchester   113

Laurence Pomeroy   113

Sir Alexander Rendel   128

Halsey Ricardo   128

Harry Ricardo   128

Beatrice Ricardo, *née* Hale   128

Model Steam Engine, *c.* 1897   129

Pumping Engine, *c.* 1902   129

Dolphin Engine, *c.* 1907   160

Triumph Ricardo Motor Cycle, *c.* 1922   160

Mk. V Tanks. Battle of St. Quentin, 29th
September 1918   161

Medium Mk. 'B' Tank at Metropolitan Carriage
Wagon & Finance Company's Works   161

Dr Ormandy   164

Frank Bernard Halford   164

The B.H.P. engine, from which the Armstrong-
Siddeley Puma was developed   165

225 h.p. Tank Engines being built   192

Workshop at Walton-on-Thames, *c.* 1914   192

Workshop at Penstone, *c.* 1925   192

Sir Robert Waley Cohen   193

Sir Henry Royce   193

Tizard as a member of the Experimental
Flight, Upavon, 1915   193

Mervyn O'Gorman   193

The Rolls-Royce Eagle VIII engine which
powered the Vickers Vimy on its Atlantic flight
in June 1919   212

The Bristol Jupiter engine   213

Harry Ricardo with Harry Horning onboard
ship, 1924   230

The Ricardo turbulent combustion chamber   231

Woodside – the house my father built   242

On *The Pearl, c.* 1928. Kate Ricardo, David
Pearce, Angela Ricardo   243

Beatrice and Harry Ricardo, 1967   243

Bridge Works, Shoreham-by-Sea, Sussex,
photographed in 1990   258

Development of the Ricardo Comet combustion
system   258

Longitudinal section of the E65 engine   259

The RR-D diesel car speed record engine with
cover removed to show sleeve-valve drive   259

Wellworthy-Ricardo WS 5 compressor   259

Transverse section of four-cylinder
reciprocating steam expander   259

Preface to Second Edition

We engineers often use names without a thought for their original owners. The Ricardo turbulent head, Ricardo research engine, Ricardo Comet combustion system that heralded the diesel as an automotive power unit are now better known than the man who was Harry Ricardo.

This then is *The Ricardo Story,* from the steam-powered bicycle of his school days to the research laboratories that still bear his name. It was told by Sir Harry Ricardo himself from the vantage point of advancing years, and looks back to the dawn of our own industry. It makes fascinating, often amusing, and always enlightening reading for engineer and student alike—values recognised by the Historical Committee of the Society of Automotive Engineers, at whose instigation this edition is published.

The original first saw light as *Memories and Machines,* published in England by Constable and Company in 1968. A limited edition of the original text was reprinted in 1990 to mark the 75th Anniversary of the founding of the company (now Ricardo Consulting Engineers Ltd).

The story has now been brought up to date by Cecil French, who joined Ricardo's as Personal Assistant to the Chief Scientist in 1952 and who retired as Managing Director at Shoreham in 1990.

This is a book about people as much as it is about machines. Harry Ricardo never lived in an ivory tower. His story abounds with names of fellow workers of those pioneering years. They are the names like Hetherington, Halford, Rowledge, less familiar now, with the passage of time and the transposition to the Atlantic's other shore. To help the reader place some of these *dramatis personae* in their historical context I have added brief biographical notes as an appendix to Sir Harry's own colorful text, which otherwise remains completely intact. It is *The Ricardo Story* in his own words.

It is a story of research and achievement going back to the

infancy of the internal combustion engine, told with charm and modesty. It is a story of disappointments, told without bitterness. It has those priceless ingredients of warmth and humour, and a charity that always gives credit to the other man, when credit is due.

Apart from the contribution by Cecil French, and the biographical notes, there are some interesting photographs and drawings that have come to light since the first edition, and which illustrate various items and personalities "pictured within." The photographs come from a number of sources, but particular thanks are due to the Institution of Mechanical Engineers and the Royal Aeronautical Society, London; Rolls-Royce plc; Cambridge University; RAE Farnborough; British Aerospace; and to Drs Kate Bertram and Camilla Bosanquet, daughters of the late Sir Harry Ricardo.

Don Goodsell

Foreword

When Harry Ricardo was first asked to write a book on the internal combustion engine more than fifty years ago, he relates that the invitation came as a surprise to him, for although almost since birth he had immersed himself in engines of all kinds, he was young and little known. He took up the invitation and a faintly derisive family complained that instead of retiring like a proper author to a study, he would write amid the hubbub of family life, but the book that resulted became a much loved classic and successive editions were to remain in print until the present day. One of the original instructions was that the book was to be written in readable English and it is interesting that T.R. Henn, writing in 1960 in his book *Science in Writing* quoted a passage from it in support of his inclusion of the author as a master of clear exposition and style among such others as Newton, Gilbert White and Darwin.

It was not until 1930 that I was old enough to become aware of the author, by which time he was, I suppose, a distinguished though not yet a famous man. Of that I was unaware, but I knew that he was an ideal uncle. He had always a genius for children, an endless supply of stories, tall and very tall, and his incessant passion for making things meant that some project was always on hand whether it was the question of a model car or boat to amuse the very young, or the construction of a floating greenhouse to test a theory of market gardening. The Ricardo household was characterised by a tremendous capacity for enjoyment, and my cousins grew up amidst a range of exciting activities punctuated at intervals by extensive travel to distant parts.

This book recounts the episodes that the author considers most significant both as regards his own full life and the technical developments which he has seen and influenced. It is of interest that the year of his birth (1885) was also the birth of the first real motor cars, for in that year both Daimler and Benz in Germany first made road vehicles fitted with internal combustion engines.

The episodes are recounted very modestly, and in describing his career at Cambridge the author attributes a great deal to the then Professor of Mechanical Sciences Bertram Hopkinson, for his teaching and, above all, for his encouragement and invitation to assist in his research programme. Hopkinson's ideas were well ahead of his time, but he made an apt choice of assistant in the undergraduate who, before the age of twenty, seemed to have absorbed most of the current knowledge of the internal combustion engine, and who was at that time designing a two-stroke engine to embody an obscure theory put forward by Dugald Clerk. The engine, known as the 'Dolphin,' is covered by a patent specification dated 1906 (in which the author is described as a student) and proved so successful that it was made under license by a number of firms until the outbreak of war extinguished the market for it.

Before Cambridge came school, and the account of schooldays given here, while resembling those of others writing from a similar background, is particularly interesting in the picture it paints of the prospects for anyone wanting to learn anything of science or technology. In one of the best schools in a country grown rich largely because of its lead in technical development, the official timetable scarcely troubled about such things. Fortunately it was a tolerant establishment that allowed the author to develop his favourite pastime of finding out for himself, but clearly a good deal of determination was needed to do so.

In the First World War the author implies that it was only by chance that he was able to make any worthwhile contribution after two miserable years of inactivity. In fact it was a case of opportunity offered to a well-prepared mind. The authorities were slow to recognise the significance of the internal combustion engine in warfare, but when the need for engines suddenly became insistent both for new developments such as the Tanks, and for the disregarded aeroplane, it was essential to mobilise all the talent that this country afforded. Among the small band of people who knew what they were doing, Ricardo's knowledge and experience commanded respect, and it was for this reason that the chance was given to him. The Tanks were to play a large part in bringing the war to an end, but at their birth there was no engine of adequate power available for them, and their whole future depended upon the rapid development of an engine to meet their peculiar requirements. This challenge the author was able to meet, but it was by good judgment rather than by luck.

The start and early days of the company that is now known all around the world as 'Ricardo's' is recounted modestly enough toward the end of the book. It must have seemed an unlikely venture to succeed and must rank as one of the very few enterprises that have been able to make research and development pay without benefit of grants or subsidies. There were, of course, hard times in the early days, but many who joined the firm then were to remain for the whole of their working lives; the founder was able to provide not only the necessary technical leadership but a certain buoyancy of spirit that is so apparent in the pages of this book. He made it fun. By the time of the Second World War recognition was widespread, and honours, medals, doctorates and acclaim were to come freely from this country and abroad.

A great deal of work has been carried out in the laboratories at Shoreham-by-Sea in the fifty years' existence of the firm, but probably no single project has been of greater significance than the initial programme to investigate knock in the petrol engine, supported by the Shell Petroleum Company in 1919. In what now seems a remarkably short time the programme had more than fulfilled its objects, and provided the means for great advances in fuel quality to be made. That the programme was so quickly fulfilled was again due to the fact that the author had already formed a clear picture of the probable mechanisms of knock from his own early work. By the end of the First World War others had tumbled to the significance of knock and had also begun to research into it, but when in 1913 Ricardo persuaded a consulting chemist to supply him with samples of different fuels in order to assess them on his own test engine in the garden at Walton-on-Thames he was probably the first person ever to attack the problem methodically.

The success of the programme gave the new company a flying start, and the importance and integrity of the work carried out in the first ten years were recognised by a Fellowship of the Royal Society in 1929, a rare distinction for an engineer.

Because the author felt that radical developments in the early days of the internal combustion engine were of more interest than the steady progress of recent times and because of his awareness of the influence of his own early activities on his career, the emphasis of the book is on the early days and it ends at a time before the Second World War. This of course was not really the end of the story because with the author as chairman until 1965 and subsequently president, his company continued to prosper and expand on the lines originally designed. Further recognition was to come with a knighthood in 1948 and since that time the company has become the leading independent organisation for internal combustion engine work in the world. It serves as consultants to engine manufacturers in almost every industrialised country and its patents are used worldwide.

In writing this book my uncle had to develop a new technique. As far as I know he had never previously dictated a sentence, and all his earlier writings, whether letters, papers, speeches or books were written out in longhand, but by this time failing sight had made it necessary to resort to dictation. The transcriptions could only be read back after an interval, and thinking that clumsiness and repetition might occur with such a method, he asked me to help him with polishing the text. In fact, I have had little to do, surely a tribute to that ability for complete concentration which was apparent when writing his first book, and which like his astonishing memory has always seemed a most striking characteristic to anyone who has known him well.

Martin Howarth

Hazel Grove 1968

Harry Ricardo, aged about six

(a)

(b)

Harry Ricardo, in his early twenties (a) greasing rear wheel bearings,
(b) enjoying his pipe aboard motor launch on upper Thames

CHAPTER 1

Early Life

I was born on 26th January 1885 in my parents' London house, 13 Bedford Square, at that time a purely residential quarter inhabited for the most part by professional men, actors, barristers, architects and doctors, many of whom, like my architect father, carried out their professional work in their own homes. I was the first and, for the next four years, the only child.

My only recollections of those first four years before my sister Anna was born are of my father on a stretcher, being brought home with a dislocated back and many other injuries caused by a fall from a high scaffolding. For a time his life was despaired of, but eventually, after lying in plaster for many months, he made a good recovery. It was during this anxious period that my sister Anna was born. These two events made a deep impression which I can remember vividly to this day; my only other recollection of that early period is that of my nurse Mary to whom I was entrusted while my mother was occupied nursing my father and taking care of the new-born baby. I became very devoted to Mary. She was a young woman in the early twenties, very pretty, full of fun and always unfailingly cheerful. She was a great theatre-goer and used to sing to me the latest popular songs, such as 'Ta-Ra-Ra-Boom-Dee-A' and 'The Man who Broke the Bank at Monte Carlo'. On technical matters she was a mine of information often more imaginative than accurate. For example, she taught that locomotives were driven by jet propulsion by squirting steam backwards from their funnels, and she assured me that if I watched with sufficient concentration during a railway journey, I should see telegrams in their little pink envelopes scurrying along the telegraph wires. In this I think she displayed some guile for it was a very effective way of keeping me from fidgeting.

When about a year later Mary left us to get married I missed her sadly; a new nurse arrived, a middle-aged woman entirely lacking in charm or gaiety who devoted herself to my year-old sister, while I had to play second fiddle. However by this time my parents were returning to normal life and I was able to be with them once more.

Whilst I had been an only child I had not been without companionship, for the railed-in garden of the Square provided a safe playground for the children from the surrounding houses, among whom I made many friends of about my own age. My one dread was that I might get locked in all night, for the heavy iron garden gates were padlocked at dusk and the railings much too high for me to climb out.

My parents now thought it was time my education was begun; as a first step my mother used to take me to a small kindergarten school nearby, where I learned 'deportment', that is to say, to sit upright at my desk, to point my toes out when walking, to shake hands and to bow, and in single file to march around and around the room in time to some quite unrecognisable music. That was about all and at first I hated it until suddenly I fell head over heels in love with a most enchanting young lady named Muriel. She was younger than I, about four years old and in my eyes quite exquisite. So deep was my emotion that I could mention her name only in an awed whisper. I never spoke to her, but once, in one of our endless musical marches, I held her hand, and that was the nearest I ever got to declaring my devotion.

I did not stay long at the kindergarten school for my mother realised that intellectually I was getting nowhere. She therefore decided to teach me herself, and soon taught me to read and write and the elements of arithmetic. My father used to read aloud stories from Hans Andersen and the Grimms and, my favourite of all, Lear's *Book of Nonsense*; a little later he began to teach me Greek and Roman history. Thus my education continued until my tenth birthday when I went away to my first boarding-school, and found myself educationally at about the same level as most of the boys of my own age.

For a few days after leaving the kindergarten I was heartbroken over parting from Muriel, but my grief was short. Life for me was getting more exciting and interesting ever day, and I had much else to think about. At that time, and indeed from as far back as I can remember, I was thrilled by the sight of an engine, or indeed of any moving form of machinery and, above all, by the great mystery of how it was made.

In those days the streets of London were full of fascination. There was always the ubiquitous steam-roller to gaze at, and on rare occasions a fire-engine with its beautifully polished brass boiler at full blast belching showers of sparks high into the air as it dashed down the street with its horses at full gallop and the firemen ringing the heavy brass bell. The grandeur and the beauty of such a scene are still vivid in my memory.

London in the early nineties was still an industrial town and the streets, especially those to the south and east of Bedford Square, abounded in small factories. For instance, at the junction where Bedford Square debouches into Tottenham Court Road there stood on one side a large saw-mill and opposite a brewery, while nearby was the large factory of Crosse & Blackwell. Many of the shops manufactured their own wares in basement workshops, and very often on the ground floor in full view of the street. My father, as soon as his back was well enough, used to take me for walks through these streets and was as happy as I was watching crafts-men at their work, generally with very few tools, but great manual skill. My father was not mechanically minded but we met on common ground in our admiration for skilled craftsmanship of any kind.

In those days there was no general or coordinated public supply of electricity; hence the electric motor as a source of small power had not yet come into existence; its place being taken by the gas engine and the hot air engine. Many of the small workshops pos-sessed one or other of these which could be seen at work from the street. Moreover, unlike the engines of today, all the working parts were exposed to view and their movements leisurely enough for the eye to follow.

My mother used to take me for walks, generally to the Thames Embankment to watch the shipping in the river between Water-loo and Westminster Bridge. My recollection of the Thames is that it teemed with shipping. Endless streams of tugs and barges and dozens of ferry steamers, mostly crowded with passengers, plied up and down. The steamers were all paddle ships with very tall funnels which folded down on to their decks when passing under the various bridges. For one penny you could travel from Chiswick to Greenwich, or, of course, to any intermediate stopping-place. At that time many business people preferred to live by the riverside in Chelsea and beyond, but went to work in the City; for them the penny steamer was probably the quickest and, in fine weather, the pleasantest form of transport. Why, I wonder is it not used today? The south bank of the river was given up entirely to shipyards and factories, such as Peter Brotherhoods, Yarrows and other large engineering and shipbuilding firms, all of whom have long since moved out of London.

On these walks we generally went through the Lowther Arcade which ran alongside St. Martin's Church between Trafalgar Square and the Strand. The Arcade was lined on either side by small shops, almost all of them containing a wonderful display of every imaginable form of toy from dolls to model steam engines and boats. Sometimes on summer mornings we used to watch the start of the stage-coach to Brighton. This huge highly decorated vehicle, drawn by four horses exactly as in Regency days, started punctually from Trafalgar Square with much blowing of horns and cheering, and dashed off down Whitehall and over Westminster Bridge. It was, of course, even then an anachronism, for we had an excellent service of fast trains to and from Brighton which made the journey in one hour, while the coach with, I believe, three changes of horses, took just over three hours to reach the Albemarle Hotel in Brighton, where a sumptuous lunch was laid on. After lunch, weather permitting, the coach party was taken for a trip on the 'briny' in flat-bottomed beach boats popularly known, with some truth, as 'shilling sicks', after which the coach and party returned to London. I do not know for how long these coach trips continued but they were certainly going very strong in the early nineties and, though costly, were well patronised.

It was on one of my early walks with my mother that I encountered my first and most shattering disillusionment. On our way to the Embankment we passed a high-class chocolate shop which displayed in the window highly decorated boxes of chocolate creams, while in pride of place, on a glass shelf, stood an enormous block of what appeared to be solid chocolate, about the shape of, but considerably larger than, a building brick. The sheer sumptuousness of this fascinated me to such an extent that I always insisted on stopping to gaze at it, for I could hardly believe there could be so much chocolate in the world. I dreamed that one day I might own such a monumental block as this, and I planned that it should have a shelf of its own in my nursery where I could treasure it and exhibit it with pride to my friends. But one day as we passed, the shop window was being re-dressed, and my great block of chocolate was lying on its side on the floor. To my horror I saw that it was a mere empty wooden box coated with chocolate. My dream was shattered, and I was shocked that anyone could be so base as to perpetrate such a hideous fraud.

I have dealt at length with these walks with my parents because they stand in my recollection as highlights; by contrast the many walks I had to take with our new nurse and with my baby sister in her pram bored me greatly. We set out every morning at the same hour to walk to Regents Park, teaming up with another nurse with a perambulator holding an immense podgy boy a little younger than myself, but quite old enough to walk. I resented bitterly having to walk alongside my sister's pram holding on to the shaft, while this infernal boy rode at ease. In this formation we walked the length of Tottenham Court Road, thence along Euston Road and so into Regents Park where the two nurses sat down on the first seat while I wandered about aimlessly. Throughout the walk they discussed and compared their various ailments which so bored me that I used to count my steps all the way to Regents Park and back. As a rule my sister slept, or at least kept reasonably quiet most of the way, but sometimes she was in a diabolic mood. Her pram was littered with rattles, strings of beads and all sorts of woolly objects. Suddenly she would sit up and start throwing out all these oddments on to the pavement and I was cast loose to recover the wretched things from between the feet of passers-by. No sooner had I retrieved them than she would laughingly throw them out again. In later years Anna developed a very keen and pretty sense of humour, but at the age of twelve months her sense and fun had no appeal to me.

Bedford Square formed the southern apex of a large residential quarter comprising also Russell Square, Gordon Square and Tavistock Square. Its inhabitants would be classified as 'upper middle-class', that is to say, neither aristocratic nor opulent, but essentially respectable. In such a society it was customary for children to be segregated in the nursery under the charge of a trained nurse or governess until they reached school age. I was lucky in that I saw more of my parents than did most of my contemporaries, as I lunched with them in the dining-room and joined them again after nursery tea until I was hauled off to bed. Thus I hardly saw my parents' friends and acquaintances who generally came to dinner or later in the evening. My mother held an 'at home' every Tuesday afternoon from which I was excluded, but my baby sister was often brought in as an exhibit, and I gather usually disgraced herself in one way or another.

Our nursery was a large room well stocked with toys and a rocking-horse which I rode with great gusto, but my favourite amusement was building with wooden bricks. My father had had made for me in hard wood a large and very fine set of bricks, all of which were exactly one inch thick, but varying in length and width in steps of exactly one inch, and he taught me to recognise and to name these in terms of their dimensions. Thus, bricks of one size would be known as 6 x 3s and others as 9 x 2s, and so on. This was excellent training for I was very soon able to judge at a glance the dimensions in terms of inches of any object, and this facility has been of great value to me all my life.

Nursery life after the departure of Mary was not all grey for we had many tea-parties with other children from the Square and gradually I became more emancipated during the years that followed; in place of the tedious walks to Regents Park, I was turned loose in fine weather into the Square garden where I played with other children. In addition to the nursery, I was given a small room in the basement with a bench, a vice and a few simple tools which I called my workshop, and where I spent many happy hours trying to make things too ambitious for my limited skill, but learning a lot from my failures. On Sunday mornings my parents and I generally went by horse-drawn tram to the Zoo. We soon got to know some of the keepers who told me a lot about the various animals and their habits and, in some cases, allowed us to have some of the smaller and tamer mammals out of their cages to fondle and feed. As soon as Anna was old enough to walk she also joined us on these Sunday visits.

I would like to try to recall what I can of the face of London in the early nineties. Traffic in the streets was very light compared with today, and, of course, all horse-drawn apart from bicycles. The old 'penny farthing' still persisted but the new so-called 'safety' bicycle with equal-sized wheels and chain drive was rapidly superseding it, though the pneumatic tyre had not yet arrived. All vehicles except private carriages and hansom cabs had iron tyres. The roads for the most part were macadamised, but this was in process of being superseded by wood blocks overlaid by a thin layer of asphalt, which greatly reduced the noise level. A few well-to-do families owned private carriages, as did such professional men as doctors, but the bulk of the traffic was made up of tradesmen's carts and heavy drays, and four categories of public-service vehicles, the horse-drawn tram, the horse-drawn bus, the four-wheeler cab generally known as the 'Growler', and the hansom cab. Of these the tram was the cheapest and slowest, and the hansom cab the most expensive and fastest form of travel. With rubber tyres and a fast-trotting horse, one could count on an average speed of between 12-14 m.p.h. in a hansom cab through any part of London except, perhaps, the actual City, and that at any time of day. My father's and my favourite form of travel was on top of a bus, which was entirely open and unprotected from the weather, except that the seats were covered by large tarpaulin aprons which could be pulled up to one's chin in rainy weather. Seated on top of the bus one was high above all the other traffic, and could survey the scene in all directions and listen to the driver's running commentary on current events and the iniquity of politicians.

In the absence of any through traffic the broad roadway within Bedford Square was, in fine dry weather, ideal for roller-skaters and for those learning to ride a bicycle. The neighbourhood saw many wandering street vendors such as the muffin-man, bell in hand and a huge tray of muffins balanced on his head, the man wheeling a barrow and shouting 'mussels and cockles alive, alive-oh', another with a brazier offering hot-roasted chestnuts, another whom my father always patronised who came round with a portable treadle-operated grindstone and offered to sharpen all our knives, scissors, pen-knives and any other tools, which he did with great skill. His pockets, I remember, were full of match-sticks, with which he tested the sharpness of the blade, and he was not satisfied until the whole length of it was razor sharp : for him I always remember we collected our used match-sticks. There were also the entertainers, the Punch and Judy shows and the organ-grinder with his tame monkey who collected pennies thrown from windows or handed out by passers-by. There was a clown who strode about on enormously high stilts, and the one-man band with a drum strapped to his back, which he played with drum-sticks fastened to his elbows, with cymbals on his head operated by a string attached to his foot, and blowing some sort of wind instrument as well. There was plenty going on to watch from my nursery window on fine days, but on wet winter days it was very different; the garden of the Square looked dank and dismal, and the roadway was deserted.

Thus far I have described only the brighter side of life in London as it appeared to me during my nursery days; the other side to the picture was the grinding poverty and squalor of such slum areas as Seven Dials and Soho, only a few hundred yards away. My father at that time was a member of our local council, and he and his friends did their best to awaken the authorities to the need and urgency of drastic action. One evil was the appal-ling amount of drunkenness. In the streets of the area drunken brawls were everyday occurrences, and it was considered unsafe for a woman to go out after dark without a male escort. Street lighting was by gas and very inadequate. Although the larger thoroughfares such as Holborn and Tottenham Court Road were lighted by lamp-posts at fairly frequent intervals and, to some extent, by the oil lamps of the vehicles, both side streets and residential backwaters were in almost complete darkness except for the pools of light immediately below each of the widely spaced lamp-posts.

Gas was used for lighting and power production, but never for space heating, for which coal was the only available fuel, and therefore smoke issued from the chimneys of nearly every house. Small wonder that during the winter months dense fogs were frequent, and sometimes so persistent as to last for three or four days and nights. At their worst these fogs brought all traffic to a standstill, while a few pedestrians who ventured out of doors had to grope their way in the suffocating blanket as best they could.

Despite these drawbacks and limitations, which I accepted as a matter of course, I enjoyed my life in London, and once having learnt to read I devoured books and periodicals dealing with engines and mechanisms generally, and the adventure stories by Henty and Rider Haggard. From the earliest age I was well-versed in the art of mother baiting and at my worst behaviour in her presence. For example, on one occasion, when she took me to a large children's party at a neighbour's house and enjoined me to be on my best behaviour, I strode into the room full of mothers and children, stared round, and then announced in a loud voice 'there are too many people here and they are all ugly'.

In later years, having children and grandchildren of my own, I have come to learn that mother baiting is not so much an art as an instinct inherent in children from the moment they are born, or even before, but according to my mother's stories, all to my discredit, it was in me highly developed.

From an early age I had very decided views as to what I wanted to do or to possess, and I bullied my parents unmercifully until, worn out by exhaustion, they generally gave in and let me have my way. As time went on I gathered from experience the elements of diplomacy, and learned never to press my demands far enough to provoke a blank refusal from which my parents could not retreat without loss of face; hence, when the danger of a positive veto seemed near I would drop the subject, to re-open it on another occasion and, if possible, from a different angle.

When I was about seven years old I found among my many acquaintances in the Square gardens a kindred spirit in a boy about a year older than myself. He was an electrician and seemed to know all about electricity. We soon became great friends and until he went to a boarding-school we used to spend many hours together in my small workshop, messing about with dry batteries and minute electric-light bulbs. Our most spectacular achievement was the installation of electric light in every room in my sister's dolls-house, so that in after years, my sister used to boast that her house was, by several years, the first in Bedford Square to be lighted by electricity. So elated were we by this success that we planned to provide the whole of London with electric light. He was to look after all the electrical equipment, and I to design and build gigantic engines to operate it, a powerful combination. Unfortunately our plan never materialised, due, no doubt, to opposing vested interests.

On our regular Sunday visits to the Zoo my favourite haunt was the tropical small mammal house where my favourite animal was a kinkajou, a most endearing little creature to which I grew very attached. Being a nocturnal animal it was generally asleep when we arrived, but the friendly keeper used to take it out of its cage and let me fondle it. It generally ate a few raisins, yawned cavernously, and then cuddled up in my arms and slept. The keeper told us that it was a very rare animal found only, I think, in the tropical forests of Madagascar, and that it was the only living example the Zoo had been able to acquire. I had set my heart on having one as a pet, and needless to say I pestered my parents to get me one. In vain they argued that it was not procurable; that being a tropical creature it could not live in our cold house, and that, being nocturnal, it would not make a good pet. I brushed these arguments aside as merely frivolous and clung to the hope that by judicious persistence I should in time succeed.

In the autumn of 1893 my parents and I were staying at my grandfather's house along with a large party of other grandchildren and their respective nurses. For some days I had noticed that a strange, new and unattached nurse had appeared on the scene. One morning this nurse told me that my mother had a wonderful and lovely surprise for me. At once I concluded that the long hoped for kinkajou had materialised. With the nurse I went up to my mother's bedroom where to my surprise she was still in bed; the nurse groped under the bedclothes and hauled out a small object and said, 'Look, this is your new dear little sister —won't you be happy with her?' I burst into tears and ran out of the room; I had expected a kinkajou, but all I was being offered was an unwanted sister. Anna by then was four and a half years old, and was gradually becoming companionable, but I had suffered much during her earlier years from the wreckage she made of my toys, and her habit of knocking down my buildings, and the thought of having to go through all this again was grim indeed.

In the event it was Anna, not I, who suffered, for I had left the nursery and gone to school before Esther was old enough to do much damage. I am afraid I did not give her a very warm welcome on her arrival in this world, but I hope that in later years I made it up to her, for we became great friends and excellent companions.

After Esther's birth only a little over a year remained before I was destined to leave the nursery and go to my first boarding-school. During that period my parents were very busy, my father with his architectural work and a pottery in Fulham in the absence of his partner, de Morgan, of whom more later. My mother, who had had some training as an accountant, was occupied checking over the accounts of my grandfather's firm of consulting engineers. Hence they had little time to devote to my education, and there-fore enlisted the part-time help of a Swiss governess experienced in coaching those nearing school age.

Mademoiselle Hunsaker, who later became Anna's governess was, I think, quite a good teacher. She taught me the elements of algebra and Euclid and an outline at least of the Christian religion, about which I was abysmally ignorant. My mother had read to me stories from the Bible, but they were mixed up in my mind with other stories from history, such as those of King Alfred and Canute. I never worried much about what would become of me after death, but from odd scraps of information I picked up I understood that I should immediately turn into a 'soul'. In my literal mind I pictured the change coming about in much the same manner as I had often watched a caterpillar shed its skin and change into a chrysalis. In this form, and pro-vided I had been good, I would go to heaven and be very happy with lots of angels for playmates, but when I asked my nurse what sort of engines I should find there, she doubted if there would be any. When I asked if I should be able to have a workshop of my own she said she was pretty sure that God would not put up with the muddle and mess I made in my existing workshop. Thus I took a dim view of the promised delights of heaven. If, on the other hand, I had been very bad, I should find myself in an alternative establishment called hell, where it was very hot, the accommodation very uncomfortable and where, from all accounts, the conditions might be more favourable for steam engines. My nurse, however, assured me that in that environment I should be very unhappy, and I concluded that on balance I would prefer heaven to hell.

As part of my arithmetic lessons I 'helped' my mother with her accounts. In these she always set out two columns of figures, one headed 'credit' and the other 'debit': we then added up all the figures in each column, and then subtracted one total from the other, leaving what she called a 'balance'. I assumed that God adopted the same method over my accounts, putting my good deeds in the credit and my bad ones in the debit account, but it puzzled me how he evaluated each separate item. For example if, when my nurse was not looking, I drew patterns in treacle on the nursery tablecloth that would be accounted a bad deed, and God would record it in the debit column. If, on the other hand, in a moment of unusual generosity I gave Anna one of my sweets, that would be accounted a good deed and would be marked up in the credit column, but would the two cancel each other out, or would I have to offer Anna another sweet to balance the account? This, I felt, would be asking too much. There ought to be a price-list of all imaginable deeds, good or bad, and no doubt God had one, but I had to rely on guesswork, and for all I knew I might be deep in the red.

I did not, however, worry much about my after-life; it all seemed so vague and remote, and I was much more concerned with such immediate problems as whether it would be better to employ in the steam engine I was designing a slide rather than a piston valve.

This period marked the end of my nursery life in London. We did not, of course, spend the whole of my first ten years in town, but went for a month at least every summer to my grand-parents, or to various seaside places where my father and I loved to go sailing and fishing, and where I could indulge in my other chief hobby of butterfly collecting.

My grandfather had acquired a large country house named Rickettswood about mid-way between Horley and Dorking where he spent most of the summer months surrounded by as many members of his family as possible, together with a number of family friends. He himself went daily to his London office, leaving early in the morning and not returning until dinner time. This involved a five-mile drive in his carriage and pair to Horley station and a half-hour train journey to Victoria, from which he always walked to his office in Great George Street.

My grand-parents always enjoyed the sight, if not the sound, of their grandchildren, but large as was the house, it could not accommodate a nursery party consisting of four nurses and fourteen children and their parents, all at the same time, so we visited in relays of two families at a time, generally for three or four weeks each summer.

The Rickettswood estate was a little over 400 acres farmed by a bailiff, a wild and totally illiterate Irishman who was devoted to my grandfather and all the Rendel family. He was particularly kind to us children and always made us welcome at his farm.

Near the far end of the estate stood a large cottage which my mother's sister, Edith, had converted as a convalescent home for London slum children. My Aunt Edith, unlike her two sisters, never married; she devoted herself to social welfare and in addition to the convalescent home at Rickettswood she ran a holiday camp for London factory girls and a club for them in London, all of which she organised with great ability and the most sympathetic understanding of the girls' needs. In her work she was later helped by my cousin, Leila Rendel, the eldest grandchild, and later founder of the Caldecott Community.

The servants, and to a large extent the entire household, were ruled by Lucy, an elderly woman who had been in my grandparent's service for many years, and had gradually become the self-appointed guardian of the Rendel family, of their morals, their manners and their etiquette. She held very strong views on many aspects of our life, and insisted that even in the country all grown-ups should dress for dinner. She held, too, to the Victorian rule that children might be seen but not heard, and that the place for them was the nursery, or a part of the garden laid out as a playground with plenty of trees to climb, a large sand heap and a paddling-pool. Lucy, despite her harsh exterior, was devoted to both my grand-parents, and the Rendel family as a whole. In her eyes Rickettswood Rendels were the salt of the earth, but I don't think this adulation was given to other branches of the Rendel family. To Lucy all Rickettswood grandchildren counted as true Rendels, even though we were half-castes, but my uncles and aunts-in-law belonged to a lower order of life, as the hint of disdain in her manner towards them indicated.

My mother, who was a Rendel, was quite at home in the bosom of her family at Rickettswood but my father, being merely an 'in-law', felt rather out of place. A lover of music and of the arts, he must have regarded the Rendels, who cared for neither, as Philistines, while they probably regarded him as something of a Bohemian. All that I was certain of was that my father preferred to take his two or three weeks of summer holiday alone with his wife and family, generally at some seaside place where we could go sailing and fishing. My two sisters looked back on Rickettswood as a paradise but I have not the same nostalgic memories. Except on the rare occasion when my cousin Robin was there, I found myself the eldest male by more than two years in my group of cousins. In London I had won my release from the nursery and nursery discipline at the age of ten, but at Rickettswood I found myself relegated to the nursery again, and under the charge of two or sometimes three nurses, each with very different views on the conduct of children but generally superintending extremely dull walks along the country roads.

However, the opportunity for meeting groups of cousins was something always to be enjoyed. Often staying at the same time as we was my mother's youngest sister Connie who also had three children. All younger than I, the eldest, Mary Brinton, later to marry John Stocks, and now Baroness Stocks, was as vital and entertaining then as she is now.

On the rare occasions when my father joined us at Ricketts-wood, he and I used to go for long country walks looking for butterflies, birds or mushrooms, in defiance of snakes, poisonous toads and all the other horrors so dreaded by the town-bred nurses. Unlike my Rendel relations, my father was a real country lover and found interest and enjoyment in every inch of it.

It had long been his ambition to have a country house of his own and even before my birth he had acquired a few acres of land on the southern slope of a belt of woodland near the small village of Graffham in Sussex. Having selected a site for the house he started to lay out the garden in the form of three brick-walled terraces, and to clear an area for an orchard and paddock, but another twenty years were to elapse before he felt justified in embarking on the building of his house. In the meantime we used frequently to go for picnics on the site, while my father gloated over the prospect and with a local jobbing gardener planted fruit and flowering trees and shrubs. We all enjoyed these picnic expeditions, and for my father those were happy years of anticipation during which he spent much of his spare time putting finishing touches to the design of his house-to-be.

It was in 1904 that my parents started the building of their house and a small cottage at the entrance to the drive for a gardener and his wife, both of which were completed during the latter part of 1905. By this time the motor-car had developed into a general utility vehicle and I had persuaded my father to incorporate in his final design a garage large enough to accommodate two cars, and still leave room for me to have a small workshop.

He was delighted with the new house and found, as he had hoped, that he could carry on his architectural work from there. From then on my family spent most of the summer months in the new house at Graffham, with only occasional visits to Rickettswood, but by that time I was almost grown up.

CHAPTER 2

Family Background

According to family legend my father's ancestry can be traced back to a certain Portuguese Jew who, at some time in the fifteenth century, migrated to Ireland, where he raised a family which continued in Ireland for many generations. From Ireland the family moved to Holland at some time during the seventeenth or eighteenth centuries where they appear to have prospered, mainly as bankers concerned with international finance. So far as I can gather, they carried on a business similar to that of the Rothschild family, but on a much smaller scale. At about the time of the Napoleonic Wars, the family came to England, and settled in the West Country. The head of the family at that time was David Ricardo, who became famous as a political economist. David had a younger brother from whom I am directly descended. He was a member of the Stock Exchange for over fifty years, and was apparently very successful. His son, my grandfather, following the family tradition, was a banker in Bath, but his real interest lay in architecture. He married a Miss Halsey, a member of another Stock Exchange family, and they had five children, the eldest a girl named Mary, next my father Halsey, next Percy, then Arthur, and lastly Harry Ricardo. Shortly after Harry's arrival both parents died, leaving the five children in the care of their aunt and uncle-in-law, the Rev. Wedgwood, Vicar of the Parish of Dumbledon in the Wye Valley. The Wedgwoods already had a daughter of their own, and I gather were not at all pleased at having five orphan children foisted on to them. They did their duty by them as trustees and guardians, but that was all. From what my father told me the Wedgwoods were religious fanatics, with more interest in the horrors of hell than the delights of heaven. On Sundays, when not actually attending church services, the children were not allowed to play games or read any book except the Bible. They were constantly being scolded and

28

told that they were destined for hell and every imaginable form of torture; they must have had a hideous childhood, only the baby, Harry, coming in for any affection. Of the others, my father was the luckiest in that he was sent to a boarding-school at Twyford, and later to Rugby. My Aunt Mary, at a very early age, made a runaway marriage with an elderly merchant whose business was importing Japanese works of art. It was not a love marriage but afforded a means of escape for Mary. The couple went to live in Japan, and Mary remained there until almost the end of her life. I met her only once on her return to England shortly before her death.

My Uncle Percy ran away to Australia when only about fourteen or fifteen years old, and except for one very brief visit, never returned to England and my father never saw him again. From his rare letters and other scraps of information, we gathered that he was quite enjoying the life of a colonial pioneer, turning his hand to any job that came along, sometimes making a small fortune out of horse-breeding or sheep-farming, while at other times, when out of luck, working as a jobbing carpenter or brick-layer. Eventually he married and settled in Brisbane, then a small but rapidly growing hamlet, of which he later became the Mayor. He had one son and one daughter. His son, Ralph, was just my age and when twelve years old came to England for his education and, as I shall relate, we became great friends.

The next brother, my Uncle Arthur, after leaving school went into the Stock Exchange, thus following family tradition where, I gather, he did very well. He had a lovely house which my father had designed for him at Walton-on-Thames and was, I think, quite the most conventional and typical stockbroker ever known.

My Uncle Harry, the youngest of the family, went into the Army and was, when I first knew him, a captain in the 17th Lancers. He was tall, well-built and extremely handsome and, in full-dress uniform and mounted on a superbly well-groomed horse, cut a very magnificent figure. He, too, was rigidly correct in every-thing he did or wore, but, unlike my Uncle Arthur, he saw a great deal of the wider side of life, and travelled all over the world. He was a great raconteur; his accounts of his travels in India, China,

Japan, South America and both North and South Africa always fascinated me. He had, too, a fine sense of humour and the most polished manners. He had seen active service in India on the North-West Frontier, with Kitchener in the Sudan and, later, as a major in the Boer War. On his return from South Africa he became a member of King Edward VII's personal bodyguard, and attended the King at all ceremonial occasions. He became very attached to the King, who always invited him to Sandringham to join his shooting-parties and to play bridge with him. Shortly after the King's death, Uncle Harry met with an accident while hunting in Ireland. He was thrown from his horse and sustained a fractured skull, and apparently some damage to his brain, for although to all outward appearance he seemed to have made a good recovery, his whole personality changed completely. Though only in his forties he had turned suddenly into a melancholy and rather querulous old man who shunned society. Not a trace was left of his social success, his vivacity, or his sense of humour. Instead he remained for the rest of his life a lonely recluse. He never married and must, I imagine, have left some broken hearts.

My father, Halsey Ricardo, was completely different from his brothers and a breakaway from family tradition. An architect by profession, but an artist by instinct and inclination, he was far more interested in the aesthetic than in the constructional side of architecture. His wide circle of friends was centred in the arts and crafts; sculptors such as Sir Hamo Thornycroft, painters, wood-workers and so on. He was a great admirer of all forms of skilled craftsmanship and, when President of the Art Workers Guild, used to invite such skilled craftsmen as silversmiths and jewellers to demonstrate their craft before the members of the Guild.

He was a keen musician and a first-rate pianist; he was also the happiest man I have ever known. He enjoyed every minute of his life, had no regrets, no fears or forebodings, nor had the miseries of his childhood at his uncle's vicarage left any stain but he was determined that my sisters and myself should be shielded from the sort of religious persecution which he had endured as a child. Though not a Bohemian, he was certainly no slave to convention or routine and could not understand how his stockbroker brother, Arthur, could tolerate so humdrum and regimented an existence; they had nothing in common and seldom met.

After leaving Rugby in 1872 my father went to Italy, ostensibly to study medieval architecture. There he spent the next three years, enjoying a purely vagabond existence, in the course of which he came to love both the country and its people. Shortly after his return to England he met William de Morgan, who became his life-long friend, and later his partner in their pottery venture, and shared his love for things Italian. De Morgan was one of the most versatile of men; an inventor, a skilled chemist, an artist and, in his old age, a novelist. He and my father planned, when they could acquire the capital to do so, to establish in London a small factory for the production of ornamental glazed tiles and pots.

As an inventor, de Morgan had designed and developed a very neat form of two-speed gear for pedal-cycles which, I understand, had been adopted by one or more Italian manufacturers from whom he received royalties. As a chemist he had for several years been carrying out researches in a small pottery of his own in the quest for suitable chemical compounds as colouring matter in glazing. As an artist, he shared with my father the same discriminating taste in colour and artistic design.

It was not until 1888 that they were able to fulfil their long-cherished plan and built a small factory in Fulham in which they installed a gas engine driving a grinding mill, a puddling mill, and certain other equipment including a potter's wheel mechanically driven through the medium of a very simple and effective variable speed gear of de Morgan's design. There was also a very large kiln, fired, if I remember right, by brushwood.

Although I was only a young child, my father used sometimes to take me to the factory, where I enjoyed trying my hand at the potter's wheel, but with very little success.

Both my father and de Morgan were great admirers of William Morris and some of their designs were inspired by him. They had been fortunate in getting hold of a really first-rate foreman named Passenger, himself a highly skilled craftsman and a great enthusiast. For the next ten years the two partners carried on very happily, producing glazed tiles and pots of very high artistic merit, but without profit, due probably to a lack of business experience or expert salesmanship. At first de Morgan played the more active part, but he was suffering from some lung trouble and could not stand the London winters. He had therefore to spend more and more of his time in Florence, while my father's architectural work left him less and less time to devote to the affairs of the pottery, and they reluctantly decided to wind up the business in 1898.

I well remember at Rickettswood one of my Rendel uncles saying that the de Morgan pottery venture might well have been a great financial success had it concentrated on the economic production of a few of the more popular designs and on more dynamic salesmanship. To this my father replied to the effect that that might well be, but it would have been no fun at all, but as the venture had turned out, both he and de Morgan and a small staff of enthusiastic and enterprising employees had bought, at small price, ten years of sheer enjoyment, experimenting with new designs and new colour schemes, whereas the alternative would have been sheer drudgery.

On his return to England from Italy, my father had been articled to an architect in Oxford, and it was there that he first met and fell in love with my mother, Catherine Jane Rendel.

The Rendel family were a typical product of the Industrial Revolution, and little is known about their origin. My great-great-grandfather, a Devonshire farmer, owned a small farm near Okehampton, and was also a road surveyor. His son, James Meadows Rendel, was ambitious to become a Civil Engineer and, at a very early age, was apprenticed to Telford, then at the height of his fame in the West Country. Telford evidently thought very well of the youth, and entrusted him with a good deal of responsibility. After several years as assistant to Telford, whom he admired greatly, the young James decided to set up on his own as a Consulting Civil Engineer. This proved a great success. From a small beginning in Plymouth, the practice grew, during James' lifetime, into a vast concern embracing harbours, docks, bridges and railways. His principal achievements were harbours such as Portland, such docks as Sunderland, and the Victoria London Docks; he built the first swing bridge at Boscombe in 1826, and the floating bridges at Southampton and Portsmouth.

After sixteen years at Plymouth, in 1838 James moved to London, where he both lived and worked at 8 Great George Street, Westminster, on the site where the Institution of Civil Engineers now stands.

In the course of his business he met a young hydraulic engineer, W. G. Armstrong, later Lord Armstrong and founder of the great Elswick Naval Shipyard. Armstrong had helped him with the hydraulic equipment for his first swing bridge; thereafter he became James' greatest friend. James' practice continued to expand rapidly both at home and abroad, more especially in India where it included harbours, docks, bridges and the laying out of railways. He played an active part in the Institution of Civil Engineers, of which he became President, and his crowning ambition was satisfied when he was elected a Fellow of the Royal Society.

James had five sons, the eldest of whom, Alexander Meadows Rendel, later Sir Alexander Rendel, was my grandfather. Next came Lewis who died young; then George, then Stuart who later became Lord Rendel and finally Hamilton. It had been the intention that my grandfather should go into the Church, and that Lewis should join his father and later carry on the succession. With this end in view, my grandfather was sent to a school in Canterbury, and then to Trinity College, Cambridge, where he read both for a degree in Divinity and for the Mathematical Tripos. Before, however, he had completed his course at Cambridge, brother Lewis died at the age of twenty-one, whereupon my grandfather threw up his projected career to join his father in Lewis's place. Five years later his father, James, died, leaving my grandfather in charge of the whole large practice, which continued to expand under his able guidance. All three younger sons went to Armstrong's for their engineering training, and eventually became partners in that firm. George, in his day, was recognised as the leading authority on the propelling machinery for very large ships, such as battleships and heavy cruisers. In 1882 he left Armstrong's to take up the post of First Civil Lord of the Admiralty, a three-year appointment, after which he returned to Armstrong's, not at Elswick, but to found a new shipyard at Pozzuoli, near Naples, still known in Italy as the Armstrong Shipyard, and there he remained until his death in 1902. Stuart went first to Eton, and then Oxford where he read Law, joined Armstrong's as a business man rather than as an engineer, and spent most of his time in London negotiating contracts with the Admiralty and with various foreign powers. He is credited with having sold a complete navy to Japan and another to Italy, as well as individual battleships and cruisers to various South American States during the 1870's and 1880's when the great firm of Armstrong's was at the height of its fame, and before competition by the then new and younger firm of Vickers had begun to be felt. At some time in the late 1880's he became immersed in politics, and entered Parliament as a Liberal Member. By that time he had amassed a very large fortune, much of which he invested in property in London, in Brighton and in the South of France. While in Parliament Stuart Rendel became very attached to Gladstone as a personal friend, and one of his four daughters married Gladstone's son, which brought their relationship even closer.

Hamilton, the youngest of the five Rendel brothers became, under Lord Armstrong's tutelage, a leading authority on hydraulics, and was responsible for the design and development of hydraulic mechanisms for the operation of heavy guns, gun turrets, steering gears and other applications requiring great precision. He designed the hydraulic equipment for the Tower Bridge in London. Hamilton suffered all his life from a terrible stammer which made him very shy; he never married but lived the life of a recluse at Elswick until his death. Of the three Rendel brothers he was Lord Armstrong's special favourite, and was said to be the ablest engineer on the purely technical side. I never saw him.

My great-uncle George I met only once and, on that occasion, he won my heart, for he came to see me in the nursery at Rickettswood and asked me to show him my very childish designs of steam engines. These he discussed with me in all solemnity as though I were his designing draughtsman, which I found immensely flattering. My great-uncle Stuart I met on several occasions. I only remember him as a tall and very dignified old gentle-man.

It was the habit of my branch of the Rendel family rather to laugh at Uncle Stuart, implying that while his brothers were all distinguished engineers he was little more than a high-pressure salesman and that he had bought his title by contributing largely to Liberal Party funds and by friendship with Mr. Gladstone. This was probably quite unfair, and tainted possibly by a tinge of jealousy, for whatever my grandfather did, Uncle Stuart always contrived to go one better. For example, my grandfather bought an estate in Surrey known as Rickettswood and Uncle Stuart capped this by buying a bigger estate about ten miles off known as Hatchlands. Again, when, after the removal of the red flag restriction my grandfather bought his first car, Uncle Stuart re-acted by buying a bigger and better car, and so on.

In 1833 my grandfather had married Eliza Hobson, eldest daughter of Captain William Hobson, R.N., of Plymouth. In those days Great Britain really did 'rule the waves'. It was the practice of our Navy to send its capital ships to all parts of the world on survey and exploratory work. On such work Captain Hobson played a notable part, carrying out surveys in the Arctic and the Antarctic and on one of these expeditions he was instructed to survey the coast-line of New Zealand and generally to find out all he could about that country. At that time it was reported that the natives of New Zealand were a very wild and savage race who invariably murdered and, it was believed, ate any foreigners who dared set foot on their shores. Such at least was the popular belief. Captain Hobson, however, contrived to make friends with the Maoris, and even to be welcomed by them. On his return to England he reported favourably on the country and its people, and was appointed by Queen Victoria to be the first Governor of New Zealand. With his wife, his two young daughters and a party of English pioneers he returned to New Zealand where he set up a Constitution Incorporated in the Treaty of Waitangi (1840) in which I understand the whites and Maoris served as equal part-ners, a remarkable achievement. There he remained until the end of his life. His two daughters returned to England, the elder to marry my grandfather; the younger, my great-aunt Polly who never married, to take over his Plymouth home, a small but beautifully situated house and garden high up above the Hoe. We were all very fond of my Aunt Polly. As a child, and in my schoolboy days, I always enjoyed my visits to her in Plymouth, where she had many friends and admirers, especially in Naval circles, some of whom had served under her father on his various survey trips.

My Rendel grand-parents had eight children, five sons and three daughters, my mother being their eldest daughter. Of the five sons, the second, William, and the youngest, Harry Rendel, were trained as civil engineers and were destined to become partners, and ultimately to carry on the business of the family firm. The other three sons were not interested in engineering. Harry was always my favourite Rendel uncle and I believe much the ablest of them all. With the exception of Uncle Harry all my other uncles married, as also did my youngest aunt, Constance, and, between the lot, produced nineteen grandchildren. They were in two age groups, for two of my uncles, Arthur and Herbert, married rather late in life. The first age group, to which I belonged, totalled fourteen, and the second, nearly a generation younger, totalled five.

Uncle Harry was unfortunately generally abroad either in India or in Uganda, where he supervised the layout and construction of the Uganda Railway up to Lake Victoria Nyanza. Like my Uncle William, whom I hardly ever saw, he was working with my grandfather. In later years I learned how great an achievement was the construction of this railway through what was then hardly known country, with wide areas infested with the tsetse fly making the use of horses impossible, through other areas which the natives dare not enter because they said they were haunted by evil spirits, and with labour forces willing enough but utterly irresponsible. For this achievement he was congratulated by Winston Churchill and the Colonial Office. He alone of all my uncles, Rendel or Ricardo, was interested in the mechanical details of moving machinery as well as in civil engineering. On the other hand, my great-uncles George and Hamilton had been famous mechanical engineers in their day and between them, had been responsible for all the machinery, both steam and hydraulic, of the 'modern' warships built by Armstrong's during the latter part of the last century. It was with their ships that the Japanese gained such an overwhelming victory over the Russian Fleet at the Battle of Tsushima.

CHAPTER 3

Schooldays

My father had been educated at Rugby School and had a profound admiration for the liberal education it provided; in particular he had a great respect for one of the junior classical masters, a certain Mr. Whitelaw, not only as a teacher but as a guide, philosopher and friend, and who, at the time of my birth, had recently become a Housemaster. Within hours of my birth a place had been booked for me to go to Whitelaw House in September 1898 provided, of course, that I would be able to pass the entrance examination.

In those days it was considered axiomatic that before being plunged into the rigours of a public school, every boy should first be hardened off at a preparatory school where he would learn to live away from home and to fit into a community of his contemporaries. Among my father's many friends was the artist Burne-Jones, who had a week-end cottage at Rottingdean, near Brighton. He told my father of a new school which had recently started up there, and gave a glowing account of its amenities. On the strength of this recommendation and of a prospectus, my parents went to see the school housed in a large three-storey building, externally of forbidding aspect but very well planned internally and well-equipped. My mother was impressed by the excellent sanitary arrangements and the number of bathrooms; in short, in material comfort and hygiene it left nothing to be desired.

The school was run by two brothers, Thomas Mason the Headmaster and his considerably younger brother George, together with their two spinster sisters. George Mason was an old Rugbian, and since I was destined for Rugby, this was another point in its favour, while my mother was relieved by the Headmaster's statement that he disapproved of corporal punishment. So it came about that on my tenth birthday my mother saw me off by the school train along with many other parents and their offspring,

38

some of the latter in tears. I went off fairly happily buoyed up by the promise that in the holidays I would be free of all nursery restrictions. My first impression on arrival at the school was favourable but as soon as the first excitement and novelty had worn off, I was overwhelmed by homesickness.

Apart from the brothers and sisters Mason, the staff consisted of six assistant-masters, all but one of whom were young graduates filling in time before starting on their professional careers. They were not interested in teaching; they rather disliked small boys and were as bored as we were by their lessons. Seldom did any of them stay for more than two or three terms. Only one intended to make teaching his profession and tried to make our lessons interesting. In addition there was a Frenchman who seemed to be a permanent member of the establishment. Of him we were all terrified for upon the least provocation he would fly into a rage, when he kicked like a mule and boxed our ears viciously. As I remember him he was a tall man who wore gold-rimmed pince-nez and long-pointed shoes. Between his rages he appeared to be nursing some secret sorrow, and was always to be seen pacing slowly up and down our playground with his hands behind his back and looking gloomily at his pointed shoes. The school Matron was an elderly spinster who sat all day in a corner room on the top floor with one window overlooking our courtyard and the other the playground. From this vantage point she spied on us continually, and reported all our misdoings to the Headmaster. There was also a drill sergeant, a retired Army sergeant with long waxed moustaches, whose joint task it was to teach us gymnastics, which he did admirably, and to carry out our punishments, which consisted always either of marching up and down the playground in various formations or, if the number of delinquents was not large enough, making us roll the playground in dry weather, or do physical exercises with Indian clubs in the gymnasium for what seemed hours on end. I would far rather have endured a few strokes with a cane and have done with it.

Looking back I now realise that the real trouble with that school was that not one single member of the staff, male or female, had ever had children of their own, or had the slightest understanding of a child's mental processes. The school was far too new; I believe it had been functioning for little more than one year before my arrival, and neither the brothers, nor their sisters, had any previous experience of running a school. Our bodily needs were well looked after, we were well fed and kept warm indoors. On half holidays, weather permitting, we had organised games such as football, hockey and cricket, but in bad weather we were kept in with nothing whatever to occupy us, with the result that we quarrelled incessantly, or divided up into rival gangs to harass each other, while from time to time one or other of the masters, exasperated by the noise, would appear and order us to 'shut up'. If one went into any of the empty class-rooms to read a book or draw, one was immediately dubbed a bookworm and held up to ridicule. Whatever we did was wrong but no one in authority suggested any alternative occupation. It seemed as if a vast negative brooded over the whole establishment.

My father had done all he could to stimulate any initiative on my part and to build up my self-confidence. Rottingdean School did its best to destroy both: all I learnt was that the only way to avoid doing the wrong thing was to do nothing at all; while to indulge in any of my hobbies was to be dubbed a freak by my contemporaries, and sometimes to be held up to ridicule by one or other of the masters.

I am afraid I may have exaggerated the misery of my first term at school; there must have been some bright moments but I cannot recall a single one.

In the holidays that followed I had a wonderful time; true to the promise my parents had made I did not return to the nursery but had my meals with them and was even allowed to wander at will among the London streets during the daylight hours and to go by myself by bus to the Science Museum at South Kensington. Various relatives took me to such entertainments as the Buffalo Bill Wild West Show, with real Red Indians, some wonderful horsemanship, and a great deal of shooting.

Being now allowed to sit up for dinner I began almost for the first time to meet some of my father's wide circle of friends and visitors. My father never had afternoon tea, but my mother held court in the drawing-room with her women friends. I felt rather *de trop,* and found the meal of thin slices of bread and butter and cucumber sandwiches very unsatisfying. I therefore with great condescension joined the nursery tea-party, not as an inmate but as a distinguished visitor, a man of the world, an elder statesman and a V.I.P. Thus my first school holiday was a wonderful success, but I broke down completely when my time came to go back, and my poor mother must have had a harrowing time seeing her tearful son off at Victoria Station. Again I was miserably homesick; the worst time on each day was waking in a semi-conscious state believing that I was still in my bed at home, only to realise, as full consciousness returned, that I was back at school. As time went on, however, conditions improved; the appalling weather of that first ghastly winter had given way to lovely sunshine with long hours of daylight and, except during lessons, we could be out of doors until bedtime instead of milling about aimlessly indoors, getting on each other's and the masters' nerves.

The Headmaster appeared to spend his entire time in his study to which he called us individually for what we termed a 'Pi-jaw' on our iniquities as reported to him by the Matron popularly known as 'The Sneak'. During these interviews he would cross-examine us as to the motive behind our action, implying that it must have been something terribly sinister. At that age we were, of course, far too inarticulate to define our motive even if we had one. All he succeeded in doing was to reduce us to bewilderment and tears. At other times his 'Pi-jaws' would be on religion: on these occasions he would bring his face very close to his listener, a gesture I have always hated, and would speak in an awed whisper as though the subject was something indecent. He would tell us that it was our duty both to love and to fear God, which did not seem quite consistent, and would assure us that God was watching us day and night and could see not only through walls but right into the thoughts in our mind; hence there was no possibility of escaping his all-seeing eye, more penetrating even than that of the Matron's. I got no comfort from these religious 'Pi-jaws'.

I found that I had a common bond with one of the young assistant-masters who was a keen butterfly collector as were myself and several of the other boys. He had a fine collection and was a mine of information on the subject. On Sunday afternoons he used to take a small party of us armed with butterfly nets for long walks on the South Downs and seemed to know just when and where to look for any particular species. I grew to like our drill sergeant who, when he shed his parade-ground manner, was quite human. He taught us the morse code and how to signal it with flags, exchanging insulting messages with each other across the valley in which our school lay.

The next term conditions were improved in that during the summer holidays one of the smaller classrooms had been converted into a library well stocked with adventure stories and periodicals, such as the *Illustrated London News* and *The Strand Magazine*. In this room next to the Headmaster's study silence was enjoined and so, on wet days, it was always possible, if not to be alone, at least to escape from the crowd. I achieved no distinction whatever either in bookwork or games, and I can remember only a few isolated incidents of that period. For instance, during class we used, when the master's back was turned, to shoot paper darts at each other across the room using either a catapult or a simple piece of elastic stretched with both hands while the dart was held by our teeth. The snag about these conventional weapons was, firstly, that they required two hands to operate them and, secondly, that they could not be pre-loaded and fired by a trigger. I therefore set out to devise a weapon which would obviate these snags; it consisted basically of a miniature crossbow or pistol. The bow, about six or seven inches across, was made up of several laminations of clock spring which I had found could be bought by the yard from shops in Clerkenwell. The barrel was made of a length of split bamboo forming a semi-circular trough. The stock, to which the barrel was glued, was of wood cut out with a fret-saw to form a pistol grip. At the breech end of the barrel was the projecting stub of a nail which served the dual purpose of forming a breach-block for the projectile to rest against and for retaining the bowstring when fully stretched. This stub projected upwards to rather less than half the radius of the barrel, and was rounded off to prevent chafing the bowstring. To load the weapon all that was necessary was to stretch the bowstring until it could be hitched over the projecting stub, and then lay the projectile in the barrel with its butt resting against the stub : this, of course, was a two-handed operation but one that could be carried out on one's lap and under cover of the desk. To fire the weapon all that was necessary was with the thumb nail to raise the bowstring until it slid over the top of the stump when off it went. For obvious reasons I could not use the conventional form of V-shaped dart, but employed instead cylindrical paper rolls about the size of and half the length of a cigarette. On the whole this weapon behaved very well but there was a tendency for the projectile to jump out of its groove, causing a misfire. In operation I would sit leaning forward over my desk with my left hand turning over the pages of my book while, with my right hand, I would fire the projectile from below desk level, while I shuffled my feet to muffle the slight ping of the released bow. I was very proud of this achievement for it was the first piece of design, manufacture and development that I had so far completed entirely on my own.

Sundays started with early-morning prayers conducted by the Headmaster in the big schoolroom, but later we all filed off to the village church where we had to sit through an interminably long sermon by a dreary parson. There were two other preparatory schools in the same parish so that we, as relatively newcomers, were allotted several rows of pews at the extreme end of the church out of sight of the altar or pulpit.

To while away the time during the sermon some of my friends and I developed the sport of woodlice racing. With its hymn-books removed, the shelf running along the back of each pew provided an ideal racetrack. Each of us had his own favourite woodlouse, distinguished by a spot of coloured paint. The racing team would be brought in a matchbox, and by gently shaking this all the competitors would roll themselves into tight balls which we lined up on the starting-line; they then uncoiled themselves one after another and set off down the track. Each boy provided himself with a soft feather with which gently to tickle his favourite as it passed by, but this was an art calling for a good deal of skill and experience. A very gentle tickle would stimulate the creature to a fine burst of speed, but if overdone it would suddenly roll itself itself into a ball and sulk for several minutes. During the week our 'stable' was housed in great luxury in my caterpillar cage and fed on bread and milk.

About mid-way between the school and the village of Rottingdean lay a group of small modern houses one of which belonged to
my father's friend the artist Burne-Jones, who used occasionally to
invite me to Sunday tea. He expected no doubt that I had inherited
my father's love and appreciation of art, but I am afraid he found
me an utter Philistine. He was extremely kind but must have found
these Sunday tea-parties as embarrassing as I did. At that time my
favourite author was Rudyard Kipling, whose stories my father
also delighted in. At one of these rather sticky tea-parties we got on
to the subject of books, and Burne-Jones asked me who was my
favourite author. I replied at once that it was Rudyard Kipling.
Burne-Jones told me that Kipling was his nephew, and that he had
spent most of his life so far in India, but now he and his family were
coming here to live in one of the nearby houses. He said that when
he arrived he would introduce me and I would find that he would
be able to talk to me about engines. This was indeed a thrilling
prospect.

Kipling with his wife and children arrived at Rottingdean during the summer term of 1897, and we were introduced. He was,
as I had expected, perfectly charming and at once put me completely at ease. We talked, I remember, about butterflies, and he
asked me to take him for a walk on the Downs and to show him
where certain species were to be seen. This was flattering indeed.
On the appointed day he called at the school accompanied by his
daughter, a most attractive child several years younger than I.
Armed with butterfly nets we set off across the Downs to an area
where I knew the Dark Green Fritillary was generally to be found;
our luck was in for there were quite a number of these rare and
beautiful butterflies to be seen. During the walk Kipling talked to
me as though I were his equal in age and experience. At that age
I was very shy about expressing any opinions of my own to any
but intimate friends or relations, but Kipling had the gift of drawing people out and, to my surprise, I found myself chatting freely
about my hopes, my ambitions and airing my views on subjects
about which I really knew very little. Before the walk was over I
had told him all about the school, my hatred of the Matron and
of the Headmaster's 'Pi-jaws', about my crossbow pistol and about
the technique of woodlice racing. That first walk with Rudyard

Kipling was a wonderful experience. During the year that followed, my last at Rottingdean, I went for several more walks with
him and generally with his daughter whom he adored and for
whom he was then writing *Just So Stories*. On these walks Kipling would engage in conversation the farm-workers and shepherds
and the occasional stranger we met, and by means of a few adroit
questions, get them to talk freely about their affairs and interests.
He was a good listener and was no doubt collecting material for
the books and stories he was writing at the time. His presence in
Rottingdean made my life much more tolerable during my last
year at that school where, during the three and a half years of my
stay, conditions had improved considerably in that our time was
better organised and indoor games and sometimes entertainments
were provided for wet days. The choice of assistant-masters was becoming more selective and I was becoming hardened to the Headmaster's incessant 'Pi-jaws'. Nonetheless, I left Rottingdean in
July 1898, without any regrets except parting from Rudyard
Kipling.

Before going to Rugby I had read *Tom Brown's Schooldays*
which had given me a horrifying picture of life at Rugby. My
father had assured me that conditions had changed completely in
his days at Rugby and that I had nothing to fear, that public
floggings, like public executions, had been abolished long before
his day, and the initiation ceremonies, such as being tossed in a
blanket, were things of the past and that, in contrast to Rottingdean, I should find myself treated as a responsible person. In spite
of this reassurance, it was with some trepidation that in September
1898, I arrived for my first term.

My first impression of Robert Whitelaw was of a very small
man, rather stockily built with a large and completely bald head,
wearing spectacles with enormously thick lenses; though his appearance was not impressive, his beautiful voice and friendly smile
at once won my heart. He welcomed me and after explaining that
I should have to share a study as there were not enough single ones
to go round, he called for the man-servant to show me to my
quarters. The servant led me to a fairly large room containing a
wicker armchair, a small bookshelf and two cupboards, a large
table and two wooden chairs – quite a luxurious apartment. The

servant then left me saying 'I hope, Sir, you will be comfortable; if there is anything you want be sure and let me know'. To be addressed as Sir was something that had never happened to me before, and his simple courtesy combined with Whitelaw's kindly words and friendly smile, restored at once the first confidence that my years at Rottingdean had done so much to undermine. Here at Rugby it seemed I was going to be treated as a responsible person, and it was now up to me to behave as such.

The next morning I made for the school workshops. These consisted of a spacious carpenters' shop equipped with a number of benches and two wood-turning lathes : a separate department for metal-work equipped with two lathes, one a $3\frac{1}{2}$-inch, the other a 5-inch screw-cutting lathe with a compound slide rest, almost identical with my own; a drilling machine, a large surface plate and several metal vices. In charge of the workshops was an elderly man of the foreman type. He was a very skilled carpenter but had had little experience of metal-work, and in the years that followed we became very close friends. Just outside the workshops was the school power-house. This, I was told, was out of bounds, but through the windows I could see that it consisted of two very large single-cylinder Crossley gas engines, each driving single-phase Ferranti alternators by belt from enormously heavy flywheels. These engines were started up at dusk every day and shut down at 10 p.m. It is a tribute to their reliability that during the whole five years of my time at Rugby, I cannot recall a single instance of failure of our power electricity supply.

I found that within the school there were a great number of societies organised by the boys, and catering for every variety of taste or hobby. For example, there was a Debating Society, an Arts Society, a Musical Society with the school orchestra, an Engineering Society, an Entomological Society, a Bird-Watching Society, and so forth. They held their meetings in various classrooms on week-day evenings. There was also an Officers Training Corps to which nearly half the school belonged and which involved a certain amount of drilling and shooting on the school rifle-range. Any boy was eligible to belong to any of these societies on payment of a small subscription of, I think, 1/- per term. I at once joined the Engineering Society, of which later I became Secretary, and the Entomological Society. Whenever possible we enticed some expert to give us an account of the current state of the art.

Rugby was essentially what was termed a 'Classical' school, but long before my day the governing body had laid it down that every boy in the middle or Upper school, no matter what his eventual profession may be, must devote at least one period a week to what was termed 'Natural Science,' in order that he might acquire a nodding acquaintance at least with the natural laws. Natural Science was divided into two categories, one dealing with Heat, Light, Electricity and Magnetism, and the other Chemistry ('stinks') and Biology. Physics was quite admirably taught by an enlightened master with a gift for stimulating interest and imagination. We had no Physics Laboratory, but our Physics Master had collected together a lot of simple but well-chosen equipment with which he carried out experiments on the lecture-room table, the while pointing out their significance, and asking us to think over their impact on civilisation and on our way of life. He never bored us with precise figures or calculations which he said could wait till the time came for us to specialise. His objective was to teach us to think for ourselves rather than to fill our heads with a lot of dry facts and data of which we would probably fail to grasp the real or practical significance.

Far otherwise was it with our lessons in Chemistry. We had a well-equipped laboratory of which our Chemistry Master made little or no use. He seldom carried out any experiments in the lecture-room; instead he made us learn by heart the names and symbols and atomic weights of all the known elements, while he chalked up on the blackboard chemical formulae and equations which he told us to copy down without pointing out their use or significance. These lessons were a great bore to me. From them I gathered that all matter, whether gas, liquid or solid, was composed of different groupings of atoms, some simple, some complex, and that the atom was indivisible. Biology was better taught but our practical work consisted in dissecting a dogfish, too-long deceased, a messy and to me a loathsome procedure.

Classics – nowadays better termed as the 'Humanities' – were, on the whole, very well taught. Our lessons ranged over a wide field from the birth of civilisation to the present day, how nations rose, flourished, became top-heavy and fell: how the mood of a whole nation sometimes changed almost as suddenly as that of an individual; how various systems of government had been tried and had succeeded or failed depending on their flexibility in adjusting to changed circumstances, and so on. In this field Whitelaw was regarded as the leading authority, and most of the masters on the classical side followed his teaching methods, and so made their lessons interesting. A few, however, were crashing bores, and their lessons were tedious in the extreme. I suffered for a time under one of these, my form-master, for as soon as I moved to a higher form, he was also promoted and became my form-master for another two terms or more. In my school reports he always accused me and other members of his class as being inattentive. If I were headmaster of a school, I think I would sack any assistant-master who persistently reported that members of his class were inattentive, for this is surely a confession of failure that he himself was unfit to teach.

By and large I would describe our academic teaching as patchy; all the masters were, I think, dedicated to their profession, but only a few could be described as inspired teachers.

I very soon got into the way of life at Rugby where I spent five very happy years, but I am afraid my school record does not bear close inspection. Academically I climbed slowly but never quite reached the Sixth Form, while in sports my only claim to distinction was that I was a good long-distance runner and became a member of the Big Side or School Running Eight. In our long-distance cross-country races I generally came in second, always out-distanced by my best friend, Anthony Welsh, who put up new records for nearly all our traditional cross-country runs ranging from five to thirteen miles: only when Welsh was absent did I score an occasional win.

Although I enjoyed games I was never any good at those involving a fast-moving ball, probably because my mental and physical reactions were far too slow. At the butts I was a good shot at a stationary target but no good whatever at a disappearing target and a very poor shot with a sporting-gun.

My main interest, and the one to which my masters would say that I devoted too much of my time, was our Engineering Society.

I was lucky in that my arrival at Rugby more or less coincided with the first appearance of the motor-car in England. I was luckier still in that I was almost the only member of the Society who had handled one of those machines. The advent of the motor-car with its wide appeal brought a lot of new blood and new enthusiasm into our Society, and membership increased rapidly, while my first-hand experience brought me considerable prestige.

Thanks mainly to the popular appeal of the motor-car and to its rapid development during the years I was at Rugby, there was always plenty to discuss at our meetings. Arguments raged as to the relative merits of steam and petrol, or on the various forms of transmission, such as belt versus gear drive, and so on. Many and various were the arguments put forward with the didactic assurance of all teenagers. Among the leaders of these debates were Kyrle Willans, son of the famous inventor of the Willans central valve engine, and Tony Welsh. Willans was a diehard steam enthusiast, Welsh led the team supporting the petrol engine. I generally sat on the fence.

Willans had procured leave for some of us to be shown over the new factory of Willans & Robinson, where I made friends with a certain Mr. Templar who had charge of the erection and testing of all their engines. I told him I was trying to make a two-cylinder single-acting uniflow engine in the school workshop. He was quite interested and shortly after came to see how I was getting on and gave me some very useful advice and promised to put my engine when complete, on one of his test-beds. The upshot of this was that Templar invited me to visit the Works whenever I liked, and I spent many half-holidays wandering through the factory and watching every stage of manufacture from the foundry to the final test-bed. In particular I was fascinated by the then most modern manufacturing techniques such as grinding, honing and broaching.

On fine Sunday afternoons some of the members of the Engineering Society used to walk or bicycle to a point on the main Coventry road where, with any luck, we might watch the passing of one or more motor-cars out for a practice spin on one of the few well-maintained road surfaces. On one of these occasions we saw a light steam car glide swiftly and silently past – it was, I believe, a Locomobile. The contrast between this and the rattle and clatter of contemporary petrol cars was indeed striking and gave our steam enthusiasts a great fillip.

In due course I became Secretary, and my friend Tony Welsh President, of our Society. I was kept busy organising meetings and suchlike matters and to keep my end up at our many discussions, spent a great deal of time reading up all I could find in our Physics Library and the technical journals. Our President handled our meetings admirably. Tony Welsh was a brilliant young man, head of the School House, Captain of the School Running Eight and had just gained a Classical Scholarship to Trinity College, Cambridge, where he gained First Class Honours in both Classics and Engineering. He would no doubt have gone far but like most of my school friends, he was killed in the slaughter of the First World War.

Apart from the members of our Engineering Society I made many other friends at Rugby, among them Alan Don, who became Dean of Westminster, with whom I used to go shooting in the summer holidays on a grouse moor in Scotland.

Both Welsh and Don left in the summer of 1902, but I, being younger, stayed on until Easter 1903 and had some special coaching to enable me to pass my entrance examination to Trinity College, which I just managed to scrape through.

Apart from having been a second-best runner I achieved no distinction whatever at Rugby in either work or sport; my only claim to fame was the conversion of my pedal to a steam-driven bicycle. This rather absurd machine provoked much amusement among my contemporaries but, as I later learned, gave great anxiety to my poor Housemaster. It seemed that the Headmaster had called a masters conference to decide whether I should be allowed to use the machine at school. Opinion was divided, some masters argued that it was dangerous and set a dangerous precedent, others suggesting that to forbid its use would be to discourage enterprise and initiative. These latter won the day but it was agreed that after I left the use of self-propelled vehicles of any description would be banned.

CHAPTER 4

Mechanical Engineering at the Turn of the Century

For a mechanically minded individual like myself the turn of the century was indeed a most exciting period, for great engineering changes were taking place on land and sea, and the conquest of the air was near at hand. Then, as now, steam reigned supreme for all purposes requiring very large power production such as electric power-stations, big factories and ships, but great changes were taking place in the manner of its use. The large and stately steam engine, with, one might say, its gleaming muscles, its rhythmic breathing and unhurried dignity, a thing of beauty and romance, was rapidly becoming obsolete. Its place was being taken by the totally enclosed high-speed engine such as the Willans, Bellis & Morcom, Browett-Lindley and others, smaller and more economical no doubt, but in my eyes totally lacking in dignity or charm. A newcomer was the Parsons steam turbine, and I well remember how controversy raged as to whether or not the new-fangled turbine would, in the years to come, oust the piston engine altogether.

In my childhood days my great-uncle, George Rendel, who had been responsible for the design of huge marine engines for battleships and large cruisers, told me during our one and only meeting that, in his view, about 10,000 h.p. in any one unit would be the limit obtainable from a modern piston steam engine because :

(1) The reciprocating masses would be so heavy as to set up intolerable vibration.

(2) That it would be well nigh impossible to design a thrust block to cope with any greater horse-power.

(3) That it would be beyond the strength of the stokers to shovel enough coal into the boilers to feed any larger engine.

The next few years saw the advancement of the steam turbine, the arrival of the Michel thrust block and the use of oil fuel, and the whole picture of marine steam propulsion changed completely.

The turn of the century also saw the widespread use of electricity for lighting in all our larger cities, the current being generated by large numbers of relatively small steam power-stations, each serving a small district, but unfortunately at different voltages in the case of direct current, and different frequencies in that of alternating current. For example, in London one side of Bedford Square was lighted by the Marylebone power-station with direct current at 110 volts, produced by a battery of Willans high-speed central valve engines, each of about 500 k.w. capacity. The other side of the Square derived its current supply from a power-station near Covent Garden. Here four or five huge vertical double-acting cross compound Ferranti steam engines were installed with enormous flywheel alternators producing, as far as I remember, single-phase alternating current, with a frequency of twenty-five cycles per second. This lack of uniformity mattered little with lighting, but imposed a severe limitation on the use of electricity for power transmission. For example, an electric motor, suitable for use on our side of the Square, could not be operated on the opposite side, and the same limitation, to a greater or less degree, applied all over London and to most of the other large cities.

During the turn of the century little change had taken place in the design of steam locomotives. The use of superheat and slightly higher boiler pressures had led to the replacement of slide-valves by piston-valves, but in all essentials the locomotive had undergone little change for many years. Up to the turn of the century the railways, as the only means of rapid long-distance travel, had had a monopoly and were making large profits. Money was lavished on the care and appearance of locomotives, which were always things of beauty, with their steel parts like silver, their brass highly polished and their paintwork immaculate. The drivers were enormously proud of their charges and used to hold forth to a circle of admirers at every station on their wonderful achievements; it was the ambition of every right-minded small boy one day to become an engine-driver.

It was during my years at Rugby that I heard for the first time of a new form of steam engine invented by a German, a Professor Stumpf, and known as the 'Uniflow' engine. In this steam was admitted to the cylinder through a poppet-valve in the cylinder head, and exhausted through ports uncovered by the piston at the end of its stroke. On the return stroke, the residual steam was compressed up to boiler pressure; thus there was established a temperature gradient from one end of the cylinder to the other. It thus became possible to employ a very high ratio of expansion without running into condensation troubles, and a very high efficiency was obtained without the need for compounding. The new Uniflow engine was, I learned, being made in double-acting form by the M.A.N. Co. in Germany and by Sulzers in Switzerland, but it came just too late to prolong the life of the piston steam engine for high as was its efficiency, it could not compete with the Diesel engine or large steam turbine, both of which were, at that period, gaining ground rapidly.

I had always regarded the large and stately steam engine as the aristocrat of prime movers compared with which the internal combustion engine appeared much lower down the social scale, and the hot air engine, then widely used for small powers, lower still. Steam had been my first love, but the steam of my youth was warm and moist and kindly, whereas the steam of today, as used in our large ships and power-stations, is just as hot and dry and horrid as the working fluid of an internal combustion engine.

The turn of the century marked the heyday of the gas engine. Gas produced from coal was available over a wide area of England and much of the Continent, both for lighting and power. The gas engine had a great advantage over steam for small and medium power production in that it was more economical in fuel, and above all perhaps that it could be started from cold almost immediately, and suffered virtually no stand-by losses. It occupied less space than a steam engine and boiler of the same power, and required less attention. Small wonder, therefore, that it had already ousted the steam engine from nearly all industrial applications requiring powers of 100 h.p. or less.

The basic design of the industrial gas engine had become almost standardised by the 1880's, and took the form of an open-type, horizontal, four-cycle, single-acting engine, as developed by Otto in Germany. In this country its manufacture had been taken up first by Messrs. Crossley, and later by a host of other British firms. Almost invariably it was made in the single-cylinder form, even in powers up to about 150 h.p. The principal market in those years appeared to be for powers ranging from $\frac{1}{2}$ to about 40 or 50 h.p., all of the same basic design, differing in little else but dimensions. In the larger sizes it was customary, on the grounds of general stiffness and rigidity, to embed the cylinder barrel with its water jacket in the bedplate itself, in preference to the use of a separate cylinder overhanging the bedplate.

It is remarkable, I think, that the same general design should have persisted practically unchanged between the years 1880 and 1930, when the development of a uniform electricity supply brought its reign to a close. While before the end of the century the open-type horizontal steam engine had given place to the enclosed vertical multi-cylinder high-speed steam engine, the gas engine, throughout its fifty years of existence, had clung to the conventional layout and design of the steam engine of much earlier years. Apart from a few freak designs, I can remember only one serious attempt to break with convention. This was the high-speed enclosed vertical gas engine made by the British Westinghouse Company. This engine had tandem single-acting cylinders with valves in side pockets; in appearance it resembled the Willans steam engine and like that engine was usually coupled to an electric generator, but though it behaved well it was not a commercial success. In the conventional gas engine the combustion chamber took the form of a pocket containing the various valves and communicating with the main combustion space by a short but somewhat restricted passage. The valves were vertical with the exhaust below and the air inlet valve above, both being operated by rockers from a side camshaft. In addition, opening also into the valve pocket was the gas admission valve operated by a separate cam but under the control of the governor. In the smaller sizes of engine, a separate poppet gas admission valve was employed, but in the larger sizes it was usual to make the gas valve concentric with the air inlet valve. In all cases, gas was admitted in variable quantity during the latter part of the suction stroke, while a full charge of air was inhaled at all loads. The capacity of the valve pocket was usually about one-third that of the total clearance volume. Thus, at the end of the suction stroke, there was left a relatively rich mixture in the valve pocket, and the conditions were favourable for operation with a stratified charge.[1] The employment of a stratified charge was made possible, as later experience proved, by the wide range of burning of most of the gases available in those days.

In the smaller sizes of engine, what was termed 'hit and miss' governing was employed, that is to say, the gas admission valve was opened either fully or not at all; thus, on light loads, each active cycle would be followed by several idle cycles, when no fuel at all was admitted to the cylinder. A very large flywheel was therefore needed to provide reasonable cyclical regularity. In the smaller engines employing 'hit and miss' governing, hot tube ignition was used which consisted of a porcelain tube, the blind end of which was kept hot by a small bunsen flame. In the larger engines it was becoming customary to employ low-tension electric ignition with a mechanically operated contact breaker inside the combustion chamber. It has always surprised me that the hot tube ignition behaved as well as it undoubtedly did, for it afforded no positive control over the time of ignition.

All the smaller engines were started by turning the flywheel by hand, but compression was relieved by the use of what was termed a half compression cam on the exhaust valve. This supplementary cam opened the exhaust valve during part of the compression stroke and could be disengaged merely by shifting the roller of the rocker arm when the engine had got under way. For the larger engines it was customary to employ Lanchester's very ingenious gas starter.

While the British-built gas engines were almost all of relatively small power, the turn of the century saw, on the Continent, the development of gigantic gas engines running on blast-furnace gas. These, for the most part, were double-acting, four-cycle engines arranged usually with two horizontal cylinders in tandem, and with the piston rod of the rear cylinder extended to operate an enormous blower piston. In other cases, four double-acting cylinders were employed driving flywheel alternators. These monster engines with piston diameters up to one and a half metres develop-

[1] See Chapter 5.

ed about 2000 h.p. per cylinder at 80 r.p.m. During the first decade of this century, at least eight different firms in Germany and Belgium were wholly engaged on the construction of these enormous engines, almost all of the same general design, one exception being a double-acting, two-cycle engine made by Körting and also by Siegener. These large gas engines were masterpieces of mechanical design and manufacture; to see a whole row of them in operation was indeed an awe-inspiring sensation, and I was lost in admiration for the courage and enterprise of their designers. On a visit to Germany and Belgium in 1951 I saw a number of these great engines still in operation at the steel works of John Cockerill (now Cockerill-Ougrée) and was told that several of them had been in regular use for well over forty years, and were still running beautifully.

In areas where there was no public supply of gas, it was usual to install separate gas producers of a variety of types using coal, coke or, in some cases, gas oil as the basic fuel. The public supply was of a uniform calorific value and a fairly uniform composition, but the independent gas producers yielded gases of widely different calorific value and composition. This proved a headache to the manufacturers of the larger sizes of engine, for it became necessary to tailor the compression ratio and size and timing of the gas admission valve and, in some cases, the method of governing, to suit the fuel available.

When gas was not available the smaller sizes of engine were often converted to run on kerosene or lamp oil, the usual practice being to substitute a simple metering pump in place of the gas admission valve delivering fuel to an exhaust-heated vaporiser through which the whole of the admission air was drawn. This arrangement worked reasonably well but was, of course, open to the objection that the whole of the entering air charge, on its way through the vaporiser, was pre-heated to such a degree as to reduce greatly the weight of air inhaled. At the same time this increased the tendency to pre-ignition, and it was open to the objection also that with an exhaust-heated vaporiser, the latter became too hot on full load, causing cracking and carbonisation of the fuel, and too cold on light load to vaporise all the fuel delivered. Despite these objections, the vaporising oil engine was widely used, more especially in the Middle and Far East, where gas was not obtainable.

There were, of course, a great many attempts to get over these difficulties, the most successful of which was, I think, the engine developed by Ackroyd-Stuart, and known as the Hornsby-Ackroyd oil engine. In this engine a large closed bulb was attached to the breech end of the cylinder head. Into this the liquid oil was sprayed during the suction stroke. The bulb was uncooled and, under normal working conditions, attained a temperature high enough not only to vaporise the oil but, as in the case of the hot tube in the gas engine, to ignite the charge. The cycle of operations was as follows: at the end of the exhaust stroke the bulb was left full of residual exhaust products. During the ensuing suction stroke cold air was drawn into the cylinder in the normal way while, at the same time, kerosene was sprayed into the hot bulb where it vaporised, leaving the bulb full of a non-inflammable mixture of oil vapour and residual exhaust. During the compression stroke air from the cylinder was forced into the bulb thereby forming a combustible mixture which ignited at or about the appropriate time. To start such an engine the bulb had to be heated by a blow-lamp or other form of external heat, as in the case of other vaporising oil engines, but, once started, combustion within the bulb maintained the necessary temperature for vaporisation and ignition. Needless to say, the efficient functioning of the engine was dependent on the capacity of the hot bulb.

The Hornsby-Ackroyd was, I think, the most all-round successful oil engine of its day and was the forerunner of the so-called semi-diesel engines so widely used some twenty years later.

Another successful kerosene oil engine of about the same date was the cold-starting Priestman engine. This, like the true diesel engine, employed an air compressor to deliver direct into the cylinder during the suction stroke a very finely pulverised spray of liquid fuel in the form of a mist. This was vaporised by the heat of compression and subsequently ignited either by a hot tube or electric ignition.

The Priestman engine functioned very well; it had the merit over all its rivals that it could be started instantly from cold. It shared with the Hornsby-Ackroyd the advantage that the air charge was not pre-heated, but it suffered the drawback that the air compressor absorbed a good deal of the available power, and was itself an expensive component.

Another comparatively new development of that period was the diesel engine which could justly claim to be the most efficient form of prime-mover in existence. Since my schooldays controversy has raged as to the proportion of credit due to Diesel and to Ackroyd-Stuart for its initiation. My own view is that the lion's share should go to the very able group of engineers at the German firm of M.A.N., for it was they who made the engine work in a practical form.

One other form of industrial heat engine deserves mention, namely, the hot air engine. Originally invented and patented by the Rev. Stirling in the early years of the nineteenth century, it came into wide use during the latter part of the century for all purposes calling for very small power production of the order of $\frac{1}{2}$ h.p. or less. The engine took various forms but at that time the two most popular were the Rider type manufactured by Hayward Tyler in London, and the Stirling type made by Gardners of Manchester. The former was a very massive and clumsy design, but none the less very popular for pumping water in country houses and farms. It had the virtues of being simple, silent and vibrationless, and was usually fired by coal or wood. The Gardner engine, on the other hand, was an elegant piece of mechanical design, very light and compact. It was usually fired either by gas or kerosene, and, in my young days, was to be seen at work behind many shop windows in London grinding coffee or operating moving models.

The beginning of the twentieth century saw the final extinction of the hot air engine by its new rival the small industrial petrol engine, and in certain limited areas by the fractional horse-power electric motor, whose use was gradually being extended by the spread and growing uniformity of public electricity supplies.

During the nineteenth century it had been the practice in factories to employ large slow-running steam engines, power being transmitted usually by multiple rope drive to a series of line shafts, from which individual machines were driven by flat belting. In order to keep the length of line shaft down to the minimum, the factory was generally built in several storeys, but in the newer and more modern factories, electric transmission was being employed with a separate electric motor driving its own short length of line shafting in each individual department. This allowed single-storey buildings to be employed, thus greatly facilitating the handling of heavy material; however, the complete replacement of line shafting and belt driving by an electric motor incorporated into each individual machine had yet to come. In the new and, probably, for its day, the most up-to-date factory of Willans & Robinson at Rugby, the power-house, consisting of several high-speed Willans engines direct-coupled to dynamos was an entirely separate structure. The factory was a long single-storey building at one end of which was the foundry and forge, while the other end formed the erecting and test shop, the middle part being devoted to all the machining operations. Overhead electric travelling cranes made by Royce (later to form Rolls-Royce Ltd.) ran from end to end of the entire building, a very efficient and practical layout.

To my mind the most spectacular and far-reaching development of that period was the revolution in road transport. The railways long since had superseded the stage-coach as a means of transport, with the result that the roads, with the exception of a few main roads, had fallen into decay. Many of the secondary roads had degenerated into mere cart-tracks, with deep ruts and very rough surfaces, very dusty in dry weather and muddy in wet. The new-fangled safety bicycle had come into fashion in the early nineties, but with its thin and narrow solid rubber tyres and the absence of any springing, its use was virtually restricted to towns and a few main country roads with reasonably smooth surfaces. At some time in the middle of the nineties the pneumatic tyre made its first appearance; thus fitted, the safety bicycle could be ridden in reasonable comfort over all but the worst road surfaces. The advent of the pneumatic tyre gave a tremendous fillip to the cycle trade, which appeared to concentrate in the Coventry and Birmingham area.

The early steam coaches built by Hancock & Goldsworthy Gurney during the early years of the nineteenth century had been masterpieces of mechanical design in their day, but had been banned largely because their presence on the roads frightened horses, and partly because of sheer prejudice. Laws had been passed forbidding the use of mechanically propelled vehicles on public highways without the escort of a man walking in front with a red flag. This, in effect, restricted the use of heavy mechanically propelled vehicles to steam traction-engines and steam-rollers. No such restriction applied to other European countries or America. In both Germany and France pioneering firms such as Daimler, Benz and Panhard were developing relatively light carriages, propelled by various forms of internal combustion engine using, what was then termed, petroleum spirit as their fuel. It was not until November 1896 that the restriction in England was removed, when a number of foreign automobiles were imported and purchased by enthusiasts and were first seen on the public roads in this country. Among the first owners were my grandfather whose first car, a $3\frac{1}{2}$ h.p. German Benz, was delivered in the early summer of 1898, while, in the same year, my great-uncle Stuart acquired a French Panhard car.

Even before the restriction was removed, a few wealthy enthusiasts had imported foreign cars for use in their own private parks. Among these was the Hon. C. S. Rolls who possessed a French car.

Until the lifting of the ban, there was no incentive for any manufacturing firm to undertake the development of mechanically propelled vehicles; with its removal a new industry sprang into being in this country. While the foreign manufacturers had already had several years' experience in the design, development and manufacture of light mechanically propelled vehicles, we, in this country, had to start from scratch and grope our way as best we could.

All these early developments took place while I was still at school, but I followed them with intense interest. Until the end of the nineteenth century the motor-car remained the rich man's fascinating but wayward toy with which one played but with little hope of reaching one's destination. As a mode of transport over short distances the horse still reigned supreme. My grandfather, then in his seventies, never went out for a drive without being followed by a horse and carriage; as often as not to return in the latter leaving the chauffeur and myself to get the car home as best we could. Progress, however, during the next few years was rapid; by 1905 the motor-car had become a practical and reliable vehicle capable of making long journeys with reasonable certainty of arriving at one's destination on time, while the horse was becoming the rich man's toy. In that year my grandfather disposed of his carriage, and the stable-boy was trained as a chauffeur.

My earliest experiences were with a newly acquired Benz dog cart in the summer of 1898. I will not attempt to describe the car in much detail : briefly it was a four-seater, two in front and two behind, the latter facing backwards. It had two very large rear wheels, solid rubber tyres and two very small front wheels. The steering pillar was vertical with rack and pinion steering. In place of the steering wheel was a short lever or tiller. The only brakes consisted of pads pressing against the rear wheel rubber tyres and the geometry was such that they gripped fairly well when going forwards, but scarcely at all in reverse. To prevent the car running backwards downhill there was provided what was termed a sprag. This consisted of a heavy steel rod hinged at one end to the frame of the car; the other end, shaped like a chisel, trailed on the ground but was normally raised clear by a chain. When going uphill the sprag was left trailing on the road. The idea was that if the car threatened to back downhill, the chisel end of the sprag would dig into the road and hold it fast. We never trusted it, and when negotiating a steep hill we always shed one of our passengers to walk behind with a brick to put under the back wheel. The engine was a single-cylinder unit with all its working parts open and entirely unprotected. Its main and big-end bearings were lubricated by screw-down grease cups, and the piston by oil from a sight-feed lubricator. The engine lay under the back seat to one side of the frame with its crankshaft parallel to the rear axle. This crankshaft was extended right across the frame of the car and supported in a plummer block on the offside of the frame. From the extended crankshaft the drive was taken by flat leather belts through a countershaft provided with fast and loose pulleys by three leather belts, one of which was crossed to give a reverse.

Embodied in the layshaft were both the differential gear and an epicyclic reduction gear, known for some reason as the crypto. From the two outer ends of the countershaft the drive was taken by chains and sprockets to the rear wheels: thus were provided two forward and one reverse speed by belt drive, and a further very low speed when the crypto came into action. The normal speed of the engine was, I think, 600 r.p.m., corresponding to approximately fifteen, eight and two miles per hour. Change of gear was effected by sliding belts from the loose to the fast pulley, or vice versa, but I cannot remember how the crypto, with its very large speed reduction, was brought into operation. The carburettor consisted of a large cylindrical tank, about the shape and size of a hat box. From its lid depended a number of wicks which dipped into the liquid petrol at the bottom of the tank. Air was drawn past the wicks and so to the inlet valve of the engine. This normally provided a very rich mixture which had to be diluted by additional air on its way from the carburettor to the cylinder; hence the fuel/air ratio had to be hand-controlled under all conditions of speed and load. The fuel level in the carburettor was maintained by a hand-operated pump. There was thus plenty for the driver to do and no lack of variety, for apart from variations in speed or load, the functioning of the carburettor depended upon the volatility of the fuel and the ambient temperature, as well as upon the depth of petrol on the floor of the carburettor, and it took a good deal of skill to prevent the engine stalling from either over-richness, or weakness of the mixture strength. So far as I can remember, only what appeared to be a very inadequate wire gauze flame trap prevented a backfire into the large carburettor. However, such disastrous consequences never happened. High-tension electric ignition was provided by a trembler coil and dry battery.

The engine was fitted with a very large diameter spoked fly-wheel and to start it one had to lift the rear seat and pull the engine round by the rim of the flywheel. In practice the engine was easier to start than might have been expected, thanks perhaps to the stream of sparks provided by the trembler coil.

In retrospect I think that the mechanical design of this car was a very poor and crude example of the state of the art even at that early date; perhaps it was already a second-hand article when my grandfather bought it.

It had been our ambition to drive the whole way from Ricketts-wood to my grandfather's London house in Lancaster Gate, a distance of twenty-six miles, and we made several attempts during the autumn of 1898 but without success. It may be of interest to present-day motorists if I describe one of these trials. Our route ran through Reigate and over Reigate Hill, a long climb of about three-quarters of a mile, with an average gradient of about one in twelve, but with several short but steeper patches. This obstacle proved our undoing on each attempt. Having prepared and tuned up the car overnight, we made an early-morning start, loaded with two large watercans, a bag of tools, oil-cans and a tin of grease and several spare sparking-plugs, plenty of sandwiches and several bottles of ginger beer. Thus equipped we set out and reached Reigate, distance six miles, without any untoward incident. There we called upon an enterprising ironmonger who, with much foresight, had laid in a supply of petroleum spirit in one-gallon tins. Before topping up our fuel tank we tested several of his samples for volatility by the simple process of dipping our fingers into the liquid, and then blowing on them, and selecting the most volatile he could supply. We then set out on our attempt to ascend Reigate Hill. At the foot of the hill we stopped at a farmhouse where we drained our cooling system which, incidentally, consisted of a coil of gilled tubing in the front of the car, but without any fan. We then filled up both our cooling system and our two large watercans with cold water. We went round all the numerous exposed bearings including the camshaft and valve mechanism with an oil-can, screwed down all the grease cups, and dressed the leather driving-belts with resin. We then started on the climb.

From past experience we knew that if once we had to engage the crypto gear there would be little or no hope of getting back into a higher gear until we reached nearly level ground. We made a good start in second gear, and by careful nursing of the engine surmounted the first length of steep grade, but the next proved too much for us, and we had perforce to engage the crypto. On this absurdly low gear we could have climbed the side of a house, but our radiator was already boiling furiously, and we had lost a lot of water. We had therefore to stop and replenish the cooling system. With this very low gear in operation, vibration at any speed above about two miles per hour became quite intolerable; painfully we proceeded up the hill, my two uncles walking on ahead, our chauffeur walking alongside with his hand on the tiller, and myself following behind ready to thrust my brick under the back wheel whenever we stopped to cool down and replenish the radiator; thus, slowly and noisily, we ground our way uphill, jeered at by occasional passers-by. Long before we reached the crest our radiator had swallowed up the last drop of our reserves of water, and we reluctantly gave up the attempt for that day. We made two or three more attempts that September to conquer Reigate Hill but without success.

I was thirteen years old at that time, too young to be allowed to drive myself on the public roads, but I was allowed to drive up and down our drive at Rickettswood. Meanwhile I acted as assistant to our chauffeur, himself a first-rate mechanic and a genius at improvising. From him I learned a good deal that has been useful to me ever since.

I do not recall in that summer of '98 that we ever met another self-propelled vehicle on our country roads. We were, I am afraid, very unpopular, for our noise and appearance frightened every horse, whose driver often flicked his whip at us and shouted 'Old iron – you'll soon blow up'.

By the summer of '99 my grandfather had exchanged his $3\frac{1}{2}$ h.p. Benz for a new and much improved model by the same maker. Although basically similar, the new model had a larger engine reputed to develop 5 h.p. at 600 r.p.m. Its wheel-base had been greatly extended to give better weight distribution; it had much larger front wheels and improved steering, a much larger and more efficient radiator and self-wrapping band brakes on the rear wheels. It had also a hand brake by means of which one could lock the rear wheels. Little change had been made to the engine and transmission, the dimensions of the single cylinder had been increased, jockey pulleys had been fitted to the two forward belt drives, and the crypto gear did not give such an absurdly low ratio. Altogether it was a far better and much more manageable machine. On it we accomplished the journey from Ricketts-wood to London in one day, and that with only a few minor breakdowns on the way. With the more powerful engine and more favourable gear ratios we surmounted Reigate Hill with only a few stops to replenish our radiator.

By this summer several other cars, all of Continental make, had appeared in our neighbourhood. I well remember my great-uncle Stuart tearing up to lunch one day with a French chauffeur in his new Panhard phaeton, of which he was inordinately proud. This car was of far more modern design than our Benz. It had, if I remember rightly, a two-cylinder vertical engine under a bonnet in front, with a three-speed crash gearbox, and a large leather-faced cone clutch. It had equal-sized wheels with pneumatic tyres, and a steering-wheel in place of a tiller. The engine had a totally enclosed crank chamber, but the timing wheels and all the valve mechanism were completely exposed. Ignition was by hot tubes, each heated by a separate petrol burner. I do not remember what sort of carburettor it had, but I think it must have been an early form of spray carburettor. There was great rivalry between the two families, and still more between the two chauffeurs. We claimed that we had electric ignition, whereas my uncle had only the old-fashioned hot tube; we claimed to be able to change gear smoothly and noiselessly, as compared with his crash gearbox and heavy clutch; we claimed that with our solid rubber tyres we were immune from punctures at a time when the roads were liberally strewn with nails from horses' shoes, and thus we all tried to keep our end up but, on balance, our case was rather weak.

I cannot recall seeing any British-built cars at this time although the British Daimler Co. had started production. The first I can remember seeing was a horseless landau, an electrically propelled vehicle built by a firm of carriage-builders famous for the elegance, beautiful workmanship and finish of their vehicles. The storage batteries were concealed under the carriage floor, while the pro-pelling motors and reduction gears were concealed in the hubs of the rear wheels. Normally a two-horse vehicle, the coachman and groom sat side by side on a high box. As a self-propelled vehicle nothing was changed except that the horses were missing, and the coachman clutched a tiller instead of reins, but I cannot remember whether he retained his whip! Several of these strange vehicles were to be seen in the neighbourhood of Belgrave Square and in the fashionable church-parade in Hyde Park every Sunday, but they seem to have faded away before the end of the century.

Motor-tricycles had appeared before the end of the century, and after 1901 motor-cycles appeared in considerable numbers. For the most part these were only slightly modified pedal-bicycles, to which were attached proprietary engines clipped to the frame and driving the rear wheel by round or vee belts. So far as I can remember the most popular engine during those – my schooldays – was the $1\frac{1}{4}$ h.p. Minerva. A cycle-repairer at Rugby had one of these and in return for my help on his engine he used to lend it to me for country rides. The little air-cooled Belgian engine, like its big brother the French De Dion Bouton, set a fashion in the mechanical design of single-cylinder vehicle engines which endured for many years.

The first five years of the new century saw a period of intensive development in the design of internal combustion engines for automobile use and the emergence into fame of such great pioneers as Lanchester and Royce, both of whom I came to know well in later years. Lanchester was, I think, the greatest inventor of his day and Royce the great perfectionist. Lanchester's many and brilliant inventions, such as his completely vibrationless engine, his constant mesh epicyclic transmission gearing, his high-efficiency worm drive and much else all combined to set a standard of smoothness and silence for the rest of the industry to emulate. Royce, on the other hand, though not perhaps an original inventor, was the apostle of mechanical rectitude. He was also an artist to his finger-tips with an unerring sense of proportion.

By the year 1905 most of what may be termed freak designs had fallen by the wayside, and the general design of the automobile had crystallised into a form from which it changed very little during the next forty or fifty years, that is to say, with a four- or six-cylinder engine under a bonnet in the front, a change speed gearbox and a bevel or worm-driven live axle; it had also achieved a very high degree of reliability. There was a smaller, lighter and cheaper type of car equipped with a single-cylinder engine. This type was still popular up to the time of the First World War.

By 1905 there were reputed to be nearly 200 manufacturers of cars; most of these were small garages serving local districts, and assembling cars from proprietary component parts, such as complete engines, gearboxes, axles, etc. Of single-cylinder engines a favourite in those early days was the French De Dion Bouton, developed in France in the middle or late nineties. This engine, which set the pattern for nearly all single-cylinder cars and most motor-cycles was first taken up in this country by the Motor Manufacturing Co. of Coventry, whose output for those days must have been very large. They supplied it in either air or water-cooled form and in a range of cylinder sizes from about 1 h.p. up to 8 h.p. As an example of up-to-date mechanical design it was excellent; it embodied two flywheels, totally enclosed in an oil-tight aluminium crank chamber split into two halves along the vertical centre line. The crankshaft journals took the form of separate hardened-steel members, while a detachable crankpin united the two flywheels; thus was provided a simple form of built-up crankshaft with hardened and ground crankpin and journals running in unsplit phosphor-bronze bearings for both the journals and the crankpin. The cylinder was vertical with the exhaust valve in a side pocket, while the inlet valve was placed immediately above the exhaust, thus forming a side pocket as in gas engine practice, but with the valve pocket at the side instead of at the end of the combustion chamber. The weak point in the design of this engine was the use of an automatic inlet valve which limited its effective operating speed and was a constant source of failure. The inlet valve, complete with its seating, formed a separate assembly which could be removed and replaced in a few minutes. Woe betide the driver who set out on even the shortest journey without at least one inlet valve assembly in his spares.

After our second and improved Benz dog cart, during the next three years at Rickettswood there followed a French-built De Dion Bouton, a British-built Argyll, and a local British-built Horley car; the latter equipped with Coventry-built De Dion engines of 8 h.p. Apart from incessant inlet valve failure, these engines behaved very well and were otherwise quite reliable. For lubrication all they needed was an occasional pump full of oil to the enclosed crankcase; in appearance they were elegant, in manners clean and well-behaved, but why, oh why, did we, for so many years, endure that atrocity – the automatic inlet valve?

Throughout all this period I was still a schoolboy and had not, as yet, had any personal contact with any of the leading mechanical engineers of the day, though during one Christmas holiday, I had attended lectures at the Royal Institution by Dr. Dugald Clerk (later Sir Dugald Clerk, F.R.S.) who was then generally accepted as the leading authority on internal combustion engines. His lectures were a pattern of lucidity, and he had a gift for explaining the theory of thermodynamics in simple terms that his largely schoolboy audience could readily understand. I made up my mind that if ever I came to lecture on such subjects I would do my best to follow his example.

CHAPTER 5

Some Early Endeavours

Before describing my experiences at the university I would like to go back once again to my very early days. I have described my wanderings with my father among the many small workshops in the Tottenham Court Road area, and the delight we both took in watching really skilful craftsmen do their work. In that area almost all the skill was devoted to woodwork, such as furniture and household fittings generally. It was the skill, not the material, that fascinated me, and I longed to see the same skill applied to metalwork. It was during my first school holidays, at the age of ten, that I first explored the area of Clerkenwell and here almost all the small workshops were devoted to metalwork of various kinds. Most of them, as far as I can remember, were engaged on repairs and manufacture of clocks and clockwork mechanisms for various other purposes such as musical-boxes, and the new-fangled phonograph. Scattered among the workshops were a number of quite small foundries, factories and metal warehouses. Of these I remember best the Laystall Engineering Co., whose foundry specialised in high-class crucible castings in iron, brass, gun-metal and aluminium, and the great warehouse of Smith & Son, which apparently supplied the whole area not only with every imaginable metal in sheets, bars, tubes and other sections, but also furnished component parts for clocks, such as reels of clock springs ranging from the tiniest hair-prings to the heaviest gauge, and every imaginable form of gearing, such as bevels, worm and wheel, rack and pinion, ratchet wheels and plain spur gearing.

I loved watching the craftsmen at work, in particular in one small shop where, behind the window, sat an old man working on a very ancient metal lathe. Seeing my face glued to his window, he invited me in and thereafter I visited him again and again during my first school holiday. He was the most highly-skilled craftsman I have ever watched. There seemed almost nothing he could not tackle on that ramshackle old lathe without even a slide rest. He seemed to like my company; at least I could help him treadle his lathe, and I certainly learned a great deal from him. His turning tools he fashioned from worn-out files, and his small drills from short lengths of piano wire. After flattening the end and filing it to the appropriate diamond shape, he would heat the cutting end to a bright red heat in a bunsen flame and then plunge it into a raw potato or apple to both harden and temper it. He taught me how to solder both with a copper soldering iron and with a mouth blow pipe, such as silver-smiths and jewellers use. He was, I think, flattered by my obvious admiration for his skill and ingenuity, and it was he who first introduced me to the periodical *The English Mechanic* which became my favourite reading throughout my schooldays. This paper was dedicated not so much to amateurs as to practical craftsmen, and was a mine of information on how to tackle difficult jobs with very simple tools. The proprietors also published a kind of encyclopædia of workshop practice which was for many years, my bible. From a scientific point of view it was not perhaps very profound, but it was compounded of nearly a whole century of hard-won practical experience.

From Christmas presents and pocket money I had collected a few metal tools such as a hand-drill, a hacksaw, files, etc, and with these I tried to emulate my old mentor from the Gray's Inn Road, Clerkenwell. I soon became quite skilful at soldering and had fabricated simple structures from sheet brass or copper by riveting or brazing, and thus learned much about the characteristics of the 'seal' of various metals. I soon learned that pure copper or pure aluminium would clog the teeth of my hacksaw or files, and some materials would work-harden and require frequent annealing, and so on. From my collection of assorted gearwheels I made up long trains of gears giving huge speed reductions or multiplication, and so on.

At other times I tried my hand at foundry practice, making moulds in plaster of paris and preparing my own low melting-point alloys of tin, cadmium, bismuth and lead, following the recipes given in *The English Mechanic*. From early attempts I learned much about how to vent my moulds; how some of the alloys would contract on solidifying, while others, such as type metal, would not.

It had long been my ambition to make a real working steam-engine (not one of those toy affairs with an oscillating cylinder which I despised), but for this I had to have a lathe and to learn how to use it. I therefore pestered my parents to get me one. My uncle, Arthur Ricardo, had in his workshop a fine 5-inch screw-cutting lathe with a compound slide rest which he never seemed to use. My father bought this from him and had it installed in my small basement workshop in time for my first Christmas holiday from Rottingdean School; a wonderful surprise to come home to.

In the journal *The Model Engineer* several firms advertised complete sets of castings and material for the construction of scale model engines and with the confidence and ignorance of youth, I bought a set of castings for a one-twelfth scale single-cylinder vertical double-acting steam engine.

I had endless difficulties in making the engine and on its first test it refused to move while steam escaped from every joint. Eventually, and by dint of using very thin paper gaskets soaked in boiled linseed oil, I got the cylinder covers and valve chest joints reasonably steam-tight, but owing to faulty alignment, I had to slack off both the piston and valve rod glands till they no longer served any useful purpose as seals. At last after tears of sheer frustration came the thrilling moment. The engine worked and continued to run by itself without the aid of my finger on the flywheel. Jerkily and creakily it ran amid a cloud of escaping steam, and that was the first prime-mover I ever made.

After that first experience I vowed that I would never again attempt to make a true-to-scale model engine. I realised that I had been foolish to start off on a job requiring experience, skill and patience, none of which I possessed, but the lesson taught me the vital importance of ensuring correct alignment of all the moving parts, and of cutting down both friction and leakage to the minimum. There and then I decided that I must set to work to design a small steam engine to suit my limited skill and not to care whether it looked like a real engine or not.

I do not remember how long it took me to design and develop

this little engine, nor how many intervening phases it went through, but I was nearly twelve years old before it was completed. It was an untidy, misshapen-looking affair, and bore no resemblance to any full-sized steam engine, but it ran very well indeed, at least to the extent that it would run very fast on quite a low steam pressure, with no apparent external leakage, and developed power enough to drive my various models on the small ration of steam provided by my boiler; its obvious defects were that it was very messy in operation, spraying a mixture of oil and condensate all over the place. Also, being entirely unbalanced, it vibrated excessively at high speeds, and so brought home to me the importance of dynamic balance. Thus, at an early and receptive age, I had learned by my mistakes more about the problems of design and production than I could have gleaned at second-hand from any text-books or teaching.

I had already made a boiler for my scale model steam engine but it did not supply nearly enough steam to keep pace with the leakage and friction losses in that engine and it was only by storing up a good head of steam that I could get it to limp round for a few dozen revolutions. With my new engine, however, the boiler provided all the steam I needed to keep the engine running continuously at high speed.

At some time during the year 1897 my cousin Ralph Ricardo, arrived from Australia. His mother, my Uncle Percy's Australian wife, had died several years earlier and since then he and his father had been leading the life of typical colonial pioneers, turning their hands to any job that came along from carpentry and brick-laying to farming, but beyond being able to read and write, Ralph had had no academic education whatever. He brought with him a letter to my father asking him to arrange for Ralph to go to school in England, a rather tall order under the circumstances. My father did, however, get him into a preparatory school in Woking which catered for boys whose early education had been omitted, and to spend his school holidays with us.

Ralph was almost exactly the same age as myself, and we soon became boon companions. He entered at once into our way of life at Bedford Square, and for the next six or seven years we were virtually brothers. He was always unfailingly cheerful and happy and ready to turn his hand to anything. Like my father he had no fears or forebodings, but took life as it came and never worried about the future. Unlike my father, he had a highly developed commercial instinct, and was full of schemes for making a fortune. His colonial upbringing had taught him how to make bricks without straw; in many ways he was very ingenious, and a past-master at improvising.

At once he joined me in our basement workshop and was fascinated by my little steam engine. Straight away he urged that we should go into partnership, and produce vast numbers of such engines which we would sell at a high price and so make our fortunes. We agreed, however, that the immediate next step for the engine was to improve its manners and appearance, both of which were deplorable. In one of the technical papers we had seen an illustration of a single-cylinder petrol engine made in France by De Dion and this design appealed to us both as very clean and elegant. We decided to emulate it as far as possible. In all essentials my basic design remained unchanged except that I did what I could to lighten both the piston and connecting-rod. Our first engine, built to the revised design, looked elegant and behaved very well.

Ralph became much more dextrous than I, and though the technique was new to him, he very soon learned the art of soldering and fitting, while I did the lathe work. Together we made a good team. With his customary optimism, Ralph was sure that a glittering commercial success awaited us. Accordingly we saved up our pocket money and bought castings and material for a dozen engines. With practice and experience we found that we could complete an engine, from start to finish, in just under two hours, and we did in fact complete about seven or eight. Ralph did his best to persuade a nearby shop selling model steam engines to sell them for us, but without success. The proprietor of the shop told us that none of his customers would buy a steam engine that did not look in the least like a steam engine. After this rebuff we abandoned the project without reluctance, for we were both getting thoroughly bored with mere repetition work and Ralph was exploring other short cuts to fortune.

After seeing the possibilities of the school workshop at Rugby,

I was fired once again with the ambition to design and construct another steam engine, larger and more ambitious than my first attempt, and by this time I was nearly fourteen, and was fairly well-versed in the theory of both the steam and internal combustion engine. I have referred earlier to the Uniflow engine as the latest development, and also the swan song of the piston steam engine. Descriptions and drawings of this form of engine were appearing in the technical press and another novelty of that date was the flash steam boiler which could provide high pressure, highly superheated steam. The Uniflow steam engine employed a poppet inlet valve, somewhat similar to that used in gas engines, and unlike most valve mechanisms this valve could stand up to a very high superheat. I concluded, therefore, that a single-acting Uniflow engine and flash boiler would form an ideal combination. The engine I designed had two vertical cylinders mounted on an aluminium crank chamber and to the best of my recollection the piston diameter was $1\frac{1}{4}$ inches and the stroke about $1\frac{1}{2}$ inches. Once I had overcome the poppet valve mechanism difficulty, the only other stumbling-block was that of the piston rings, and try as I would I could not get these to give a satisfactory seal. When completed I took the engine to the test-shop at Willans' works where we ran it on steam for the first time and after a little adjustment, it ran, on the whole, very well, but there was a good deal of blow-by past the pistons. I told Mr. Templar, who was in charge of the test-shop, of the difficulty I had had with these, and he very kindly offered to true up the piston ring grooves, and get the machine-shop to make new rings for me.

This was the engine that some two years later I fitted to my pedal-bicycle, but I do not recall that I had any such intention when I started on it well before the end of the century.

My next step was to try to make a flash boiler on much the same lines as the French Serpollet used in the Paris trams, and in some private cars. My attempt was an utter failure. I discussed this with our Science Master who pointed out that such a boiler containing virtually no water depended for its thermal storage on steel tubing, whose specific heat was very low. It depended, therefore, upon the maintenance at all times of a nice balance between heat and water input, and that this would be almost impossible to achieve in the case of such a small boiler as mine. Reluctantly I decided to go for a conventional type of vertical fire-tube boiler, with as large a water capacity as I could provide.

At Rugby my time was so fully occupied in other ways that the making of this engine and boiler, despite plenty of help from my school friends, was not completed until nearly the end of my time there.

While my engine was being tested at Willans' works Mr. Templar urged me to exhibit it at a forthcoming exhibition of model engines at the Horticultural Hall in London during the Easter holidays, and this I did. The exhibition was a very fine one. Most of the contributors were professional, but one room was set aside for models made by amateurs, most of which were true-to-scale models of large steam engines and locomotives, exquisitely made and beautifully finished. Among them my little engine looked drab and amateurish, and attracted little or no interest, except for one middle-aged man who was examining it very closely. We got into conversation and he told me that he was Managing Director of a small foundry in Liverpool known as the Liverpool Casting Company, which specialised in the production of high-class small castings in iron, gun-metal and aluminium. He said that his firm sold sets of castings for several varieties of scale-model engines such as I had bought and come to grief over several years before. I told him of my sad experience and of my determination never again to attempt a true-to-scale model, which I had found far too difficult for my limited skill and patience. I mentioned that I had set out to design a form of engine with no gasket joints, no small studs and nuts, using only relatively large screws, and as few of them as possible. I told him about the little piston valve engines that Ralph and I had made and that how, as a next step, I had become intrigued with the principle of the Uniflow engine, and had sought to make a small single-acting version of this; of the difficulty I had found in machining poppet valves with very long slender stems and of ensuring concentricity between the valve stem and seat, and how I had overcome this. He seemed really interested in what I had to say, and invited me to lunch with him next day, and to bring such drawings as I had of my little engine. I was rather ashamed of the very scrappy drawings I produced, but they sufficed to give him the information he wanted. He then said that he would like to market castings for my engine, which he agreed would be much easier for an amateur to make, and although there were no patents involved, he felt that I would be entitled to at least a small royalty for the use of my design, and that he would like the engine to be called by my name. He went on to say that he thought that the selling-price of a set of unmachined castings would be about twenty shillings, and he suggested a royalty of $2\frac{1}{2}$ per cent, or sixpence on each set. He then asked me to let him have the engine for a few days after the exhibition was over.

I was, of course, delighted, and it was during my last term at Rugby that I received a note from the Liverpool Casting Company enclosing a postal order for twelve shillings, being the royalties paid on the first twenty-four sets sold. It was with great pride that I showed this to Mr. Whitelaw, to our Physics Master and to some of my school friends. It was the first royalty I had ever earned, and for several years after this I continued to receive a trickle of royalties from the Liverpool Casting Company.

During my last year at Rugby I amused myself and my friends by mounting my Uniflow engine and boiler on my pedal-bicycle. The boiler was lashed to the steering head and extended above the handle-bars. The engine I mounted rather crazily just behind the pedal bracket, the drive being by means of a roller about 3 inches in diameter, bearing against the tyre at the back wheel and projecting out sideways. This meant that I had to remove the left-hand pedal and crank. As a stationary set I had fired the boiler with a paraffin primus burner, but it was difficult to do this in position on the bicycle. I therefore resorted to coal firing. This absurd machine caused vast amusement among my contemporaries. It took a long time and much use of the bellows to get the coal fire going well, but once I had succeeded in getting up a good head of steam, I could do quite a good burst of speed for between 100 and 200 yards, after which I had to dismount to get to work again with the bellows. I had no feed pump for the boiler, but it had a water capacity of about 1 gallon, and I could carry enough coal in my pocket. As a next step I fitted a funnel to the boiler about three feet high to give some much needed natural draught, but as there was no easy way of supporting the funnel, it often fell off into the road.

By this time, the winter of 1902–3, there were already quite a number of petrol-driven motor-bicycles on the roads. The proprietor of a cycle shop in the town possessed one which he used to lend to me in return for some repairs I had carried out to his engine in the school workshop. I had been many long rides on it so I was under no illusion that my steamer could ever be a practical utility vehicle. I had rigged it up as a joke, and a joke it remained. I was constantly being asked to give demonstration runs, which always caused great amusement when, as so often happened, my tall funnel fell off.

In Chapter 2 I mentioned that it was my father's ambition to build a country house at Graffham, but one of the problems was that of water supply. A well had been dug on a convenient position on the site, but it had been driven to nearly 100 feet before reaching the water-table. This meant a total head of over 120 feet to a water tank near the top of the house, and something more than manpower would be needed to provide an adequate supply. A deep well pump made by Hayward Tyler was installed, and Ralph and I were commissioned to provide an engine to drive it. We could, of course, as at Rickettswood, have installed a hot air engine, but I was keen to design and build a petrol engine for the purpose. This was too big a job for us to tackle either in my workshop or at Rugby, but we found that during the holiday we could, for a small fee, have the use of the very large and well-equipped machine-shop installed in the basement of the Regent Street Polytechnic. Ralph, an expert scrounger, had somehow got hold of a piston, connecting-rod and crankshaft from a disused gas engine. We therefore decided to design our engine around these parts. Our design consisted of a plain box bedplate carrying the crankshaft and two camshafts mounted in ordinary plummer blocks with the cylinder mounted on four steel columns which were further extended to carry the cylinder head; a very simple if inelegant arrangement. The cylinder head was conical, with the sparking-plug at the apex, while the inlet and exhaust valves were inclined at an included angle of about 90 degrees. In order to simplify both pattern-making and machining, both valves were seated in separate detachable cast-iron elbows. Both valves were operated through long push-rods and rockers from the two camshafts mounted on the bedplate. Ralph found a foundry to make the iron casting, and with the heavy machines in the Polytechnic workshop it did not take us long to complete the comparatively small amount of machine work required.

Ralph left school during 1902 and instead of returning to Australia, went as an apprentice to the firm of Arrol-Johnston in Dumfries, near Glasgow, and for the next few years I saw very little of him. I had therefore to carry on alone with the pumping engine which I took to Rickettswood where, with the help of my grandfather's chauffeur, it was erected in a disused shed. We provided it with a conventional spray carburettor attached to the air inlet elbow and with high-tension electric ignition from a trembler coil and dry battery and rigged up a somewhat rudimentary prony brake for rough measurements of power output.

This engine, which we had built round an existing piston and crankshaft, had a cylinder capacity of just about two litres. The duty required of it was only about $\frac{1}{2}$ h.p., thus it was many times larger than was necessary. This, of course, I had realised all along but I had envisaged other uses for the engine as well as pumping. After a few teething troubles we got it running quite well on light load, but on full load we found that the rather slender steel columns on which the cylinder was mounted were not stiff enough, with the result that the cylinder kicked badly at each impulse. I had also neglected the fact that without any governor there would be a danger of over-speeding the pump. All of this I ought, of course, to have foreseen, and though it was my first attempt at designing an internal combustion engine, I felt ashamed of myself.

During the previous Christmas holidays I had attended a series of lectures by Sir Dugald Clerk in one of which he had pointed out the great advantage that might be obtained by operating a petrol engine with a stratified charge as opposed to throttling a homogeneous mixture, but he admitted that he had not yet tried this. I was much impressed by his arguments and by that time knew enough to appreciate them. The intention behind the idea was to eliminate the thermo-dynamically wasteful process of throttling and to control the engine wholly by varying the quality of the mixture within the cylinder. The mixture would necessarily be rich close to the sparking plug in order to ensure ignition but successive layers of charge further from the point of ignition would by some means be made progressively weaker; hence the term 'stratified'. If into a cylinder of air or weak mixture it was possible to insert at every cycle a paper bag containing a combustible mixture and also an igniter, then on ignition the bag would burst and its contents, already aflame, would spread throughout its surroundings to complete combustion. By varying the strength of the mixture in the cylinder as opposed to the paper bag, it might be possible to control the engine over the whole speed range.

Unfortunately, the promise of the system has never been fulfilled in petrol engines due primarily to the narrow burning range of volatile hydrocarbon fuels, for both at the weak and the rich ends of the burning range combustion is too slow for a running engine. In more than sixty years since my first enthusiasm, I have made several more attempts and witnessed many more by others, but a successful stratified charge system has not yet been achieved in a petrol engine.

However, in 1902, it seemed to me that my engine, with its conical combustion chamber and central ignition point, would be ideally suited to stratified charge operation, and I was determined to try it. I therefore replaced the sparking-plug with a small bulb to which I fitted an automatic inlet valve and sparking-plug. I retarded the opening of my main inlet valve so that during the first few degrees of the suction stroke a rich fuel/air mixture would be inhaled into the bulb and thereafter a full charge of air would enter through the mechanically operated inlet valve. Fuel supply to the bulb was by gravity from the petrol tank through a needle valve which dribbled petrol on to a wad of wire gauze at the entry to the automatic valve. This arrangement eventually worked perfectly and, incidentally, provided an automatic speed control since the rate of fuel supply was entirely independent of engine speed. Thus fitted the engine ran beautifully, smoothly and quietly over the range of b.m.e.p. from idling to about 15 to 20 pounds per square inch, which was all that I required.

After this experience I became an enthusiastic advocate of

stratified charge operation, for so far it had worked like a charm. It was not until later that I discovered its limitations.

In 1904 work started on the building of our new country house at Graffham which included the erection of a small engine house over the well head. As soon as this was completed my engine was installed and set to work in charge of an intelligent young man who became our gardener. Under his care it supplied all our domestic water needs for the next seven or eight years.

I became thoroughly ashamed of this engine as a piece of mechanical design, for the mounting of the cylinder lacked rigidity, and it was crude in other respects. That the engine had, in fact, performed so adequately was due to the very light load and low speed required of it.

After leaving Rugby I went for some months at Mr. Robinson's invitation as an honorary apprentice or assistant to Mr. Templar at Messrs. Willans and Robinson's Works. Templar proved a very good master and taught me a great deal about the testing of engines; he was extremely conscientious and methodical and would let nothing pass until he was quite satisfied. He used to set me to check and re-check the test-sheets of every engine and if I found any discrepancy, however slight, he insisted on getting to the bottom of it. Templar taught me how to take simultaneous indicator diagrams from each individual cylinder; how to measure accurately our steam consumption over the whole range of load; how to arrive at a figure for the mechanical efficiency by projecting the Willans line back to zero, and to compare this with the indicated mean pressure, as recorded on our cylinder diagrams. The two methods never quite agreed, but if they differed widely he would insist on several re-tests, and a re-calibration of our pencil indicators. From this I learned a great deal about the fallibility of indicators, and so gained experience which served me well when, a couple of years later, I assisted Hopkinson with the development of his optical indicator.

Thanks to Templar's tutelage I learned a great deal about the technique of engine testing and was forewarned of the many pitfalls that might confront the inexperienced.

CHAPTER 6

Cambridge

I went up to Cambridge in October 1903. By that time I had become a skilled mechanic with a background of practical experience of handling engines, greater probably than that of most of my contemporaries. My grandfather, however, was looking to me to join and eventually become a partner in his firm of consulting civil engineers, for I was the only grandson who possessed any aptitude for, or interest in, engineering of any kind. He urged, therefore, that I should take an Honours Degree. I explained the position to my College Tutor who directed me to attend lectures on such subjects as surveying, geology, the theory of structures and, above all, on both pure and applied mathematics. This course I followed during my first year at Cambridge, but without much success, for I could find only a tepid interest in such stationary structures as harbour works, tunnels and bridges, while the formal mathematics, on which my tutor laid so much stress, seemed to me both incomprehensible and largely irrelevant. My whole soul yearned for moving machinery, for dynamic not static structures, and it was soon borne in upon me that I would never make a good civil engineer, but I cherished the hope that when I became a responsible member and partner in the Rendel firm, I would be able to extend the range of its activities to include a department of mechanical engineering which I would take under my wing.

I was very glad once again to meet many schoolday friends including Tony Welsh, who introduced me to his circle and, in particular, to Keith Lucas, already a fellow of Trinity and a director of the Cambridge Scientific Instrument Company, who became one of my best friends. Lucas was three or four years older than I; he was, at that time, carrying out a research in biology, a research in which Adrian (later Lord Adrian) assisted him and carried on after his death in the First World War. Lucas was a genius at the design and development of delicate instrumentation. His researches into the nervous system of various animals were rewarded by his election to the Fellowship of the Royal Society and he was, I believe, the youngest man ever to have received this honour.

Of the many friends I made among my fellow undergraduates at least half were to meet their deaths in the First World War. Among those who survived it were Harry Hetherington and Oliver Thornycroft, both of whom worked with me for the next thirty years, and about whom I shall have more to say later.

At Cambridge as at Rugby there was a great number of societies, some political, some religious and others technical, but there was no engineering society as such, its place being taken by the new-born Automobile Club, whose membership was expanding rapidly, and of which eventually I became Secretary. As at school I gained much prestige from the fact that I had handled both cars and motor-cycles for a longer period and had gained more experience than other members. The moving spirits of the Automobile Club were three lecturers in Mechanical Sciences, namely, Inglis, who later succeeded Hopkinson in the Chair of Mechanical Sciences, Dykes and Rothenburg. Each of these had cars of their own. Dykes, I remember, had a Humberette with a single-cylinder De Dion engine, and was busy making in the University workshop a neat little four-cylinder engine of his own design for his car.

As soon as I saw the excellent facilities available in the University workshop I determined to make a motor-cycle of my own embodying my own ideas derived from the experience of riding many different models.

Since the beginning of the present century clock-making in Clerkenwell, which I have already described, had given place to the manufacture of component parts for motor-cycles. The firm of Chater Lea was doing a brisk local trade in the supply of frames and wheels of a rather more robust design than for ordinary pedal-cycles, and from them I bought a frame, wheels, handlebars, etc., to my own specification for a remarkably small sum. My old friends, the Laystall Engineering Co., were supplying aluminium crankcase castings of what was then the conventional De Dion design, and iron castings of air-cooled cylinder barrels and cylinder heads in a range of sizes from about 2 inches to 4 inches bore. Both the lust for speed and the spirit of competition were strong in me in those days, so I bought the largest cylinder and head castings I could get. As to the piston, I had my own ideas. The conventional practice in those days was to locate and fix the gudgeon pin in the bosses with a taper grub screw, but I did not like this. Instead I wanted to use a freely floating gudgeon pin and to make my piston as light as possible. Accordingly I made my piston in two parts, an outer member comprising the skirt and head and carrying the two piston rings which I machined all over and made as thin and light as I dared, and an inner member carrying the gudgeon pin bearing against an abutment formed in the outer helmet and screwed to it. This arrangement suited my purpose admirably as only very simple patterns were needed, and it provided a freely floating gudgeon pin safely located by the outer helmet. This form of construction had the advantage that by placing shims between the inner member and its abutment, I could raise or lower the ratio of compression.

Although by 1903 there were already many makes of motor-cycle on the market all, or almost all, were of the same general design and completely lacking in modern refinements such as clutches and change speed gear. Power was transmitted to the rear wheel by a vee-belt drive from a pulley rigidly attached to the engine crankshaft, thus the engine could not be disengaged or started independently of the machine. The usual practice was for the rider to lift the exhaust valve and pedal furiously until the engine speed was sufficient to carry it over compression. An alter-native method of starting, which I always preferred, was to run alongside and then vault into the saddle as soon as the engine got under way. In the absence of any change speed gear one had to arrive at a compromise pulley ratio, low enough to negotiate any ordinary hill, yet high enough to provide a good turn of speed on the level. On really steep hills one had either to provide pedal assistance or jump off and run alongside. Here again I preferred the latter expedient. I therefore decided to dispense with any pedal at all and fit instead long running-boards, as in modern motor-scooters, but this left me with no chance of pedalling home in the event of engine failure.

From my experience with hired machines I had suffered much from the stretch of leather belts, the only remedy for which was to remove and shorten the belt, a tiresome and messy business. By dispensing with the pedals and chain I could adjust the belt tension by merely retracting the rear wheel, a very simple operation.

To start so large a single-cylinder engine as mine with its cylinder capacity of nearly 900 c.c. and its relatively small flywheels in the usually accepted manner would have demanded a great physical effort. I decided, therefore, to follow the long-established gas engine practice of employing a half compression cam which could be brought into operation by sliding the cam follower sideways. Thus fitted I could stroll alongside the machine firing on half compression and slowly gathering speed. I could settle myself comfortably on the saddle and turn over to full compression as soon as a speed of about 10 or 12 m.p.h. had been reached. Why, I wonder, had this simple and well-known device never been applied to motor-cycle engines?

It took me less than two terms to complete my motor-cycle which, after the usual teething troubles, performed very well, but it was not as fast as I had hoped for; apart from the handicap of my automatic inlet valve which I had had to accept, the valve itself, though the largest that the cylinder head casting would accommodate, was much too small. Though the top speed of the machine was disappointing, its acceleration in the lower speed ranges was quite impressive.

I have dealt at length with my home-made motor-cycle because its success was instrumental in changing my whole future career. In my day, as now, the term 'Mechanical Sciences' in Cambridge embraced a very wide range of subjects under the overall control of Professor Hopkinson, who had recently succeeded his uncle, Professor Ewing, in the Chair. For so exalted a position Hopkinson was a very young man, in the late twenties. His ability, his personality and his charm soon won him universal respect and affection. At heart he was a mechanical engineer and, as I discovered later, the most brilliant, versatile and imaginative research leader I have ever come across.

Except on a few formal occasions I never met Hopkinson during my first two terms, but I had heard a good deal about him from

Dykes. Dykes had told me that above all else Hopkinson was keenly interested in the problem of flight by heavier-than-air machines, and that his interest had been greatly stimulated by the remarkable achievement of the Wright brothers in America a few months earlier.

It was at the beginning of my third term at Cambridge that Hopkinson invited me to his office. He told me that both Dykes and Inglis had been talking about me as a skilled mechanic, and about the success of the motor-cycle whose engine I had built in the Cambridge workshops. He asked me to tell him about my background of experience and I told him what I have recorded in previous chapters. His comment was that I had been fortunate indeed to have come in at the infancy of the petrol engine and at the birth of aviation, for both of which he predicted a tremendous future.

He next went on to ask me about my plans for the future. I told him that I was destined to join my grandfather's firm of consulting civil engineers, and hence his wish that I should concentrate on civil engineering at Cambridge. I said that my real interest lay in moving machinery rather than stationary structures, and that after my first two terms of the course, which consisted so largely of mathematics, I felt that I would never be able to take an honours degree or to become a good civil engineer. I said that I cherished the hope that I might have the opportunity to extend the range of the firm's numerous activities to embrace that of mechanical engineering.

Hopkinson listened to me most sympathetically and said that to him, as to me, mechanical engineering had far more attraction than civil, and that he would consult with Inglis and with Barnes (later Bishop of Birmingham and Fellow of the Royal Society), who was both my College Tutor and Lecturer in Mathematics. Also that he would like to see how I fared in my examinations at the end of that term.

He then turned to another subject. He told me that he personally was taking a great interest in the fuel consumption competition for motor-cycles which our University Automobile Club was organising under the inspiration of Inglis, Dykes and Rothenburg, and that he hoped that I would be one of the entrants with my home-made machine. I told him that I had every intention of competing, but that since my engine had a very much higher cylinder capacity than that of any of the competitors, I did not expect to make a good showing. He replied that, even so, that would give me all the more scope to exercise my ingenuity and urged me to do the best I could and said that he would be very interested to hear what modifications I had made to the engine and carburettor in the interests of fuel economy.

At this, my first interview with Hopkinson, I fell a victim to his charm and personality. I sensed, too, that in my case he would attach more importance to my showing in this competition, to be held in a few week's time, than to my end of term examination results. I determined, therefore, to go all out to make the best show I could.

By the beginning of my first summer term in 1904 I had the machine well tuned up and was familiar with its idiosyncrasies. Since I could provide no pedal assistance I had, in the first instance, geared it rather low. With this ratio I could negotiate any hill I had so far met with provided I was not baulked, in which case I had either to jump off and run alongside, or go back to the bottom and make a fresh start, but with so little traffic on the roads this seldom happened.

For the forthcoming fuel consumption competition the route chosen was from Cambridge to Royston, to Newmarket and back to Cambridge. The conditions were that all the competitors should assemble at the first milestone on the Trumpington Road, where their petrol tanks and carburettors would be drained and their pedal chains removed. Each would then be served with one quart of petrol and thereafter it was up to him to eke this out as best he could. On running out of fuel they would wait by the roadside until Inglis or Rothenburg arrived to record the distance they had covered and re-fuel their petrol tanks. Speed was to play no part and all competitors were advised to run at whatever speed they thought fit, the one and only criterion being the distance covered on one quart of petrol. All were free to and were urged to make any adjustments or alterations to their machine that they considered would reduce its fuel consumption.

My engine had by far the largest cylinder capacity of any of the competitors and the betting was that I should be the first to fall by the wayside. Spurred on by Hopkinson's challenge, and the knowledge that he himself would take a particular interest in my performance, I devoted most of the next three or four weeks to preparations for the event. My first step was to familiarise myself with the course. There were no gradients worth worrying about and very few sharp bends or corners to call for slowing down or acceleration. I concluded, therefore, that I could reckon on maintaining a uniform speed of about 25 m.p.h. for almost the whole of the course, and therefore need not disturb my carburettor conditions once I had got them adjusted to the optimum. Since I had an ample reserve of power and no steep gradients to face, I fitted a very much larger pulley to my engine and found that even with this high gear, I still needed only a very small throttle opening to maintain a speed of between 20 and 25 m.p.h. With this high gear ratio combined with my large cylinder and undersized flywheels, during acceleration the machine proceeded in a succession of violent lunges, and the sensation was that of riding on a kangaroo.

For fuel consumption tests I had fitted a very small pilot petrol tank holding only about one-third of a pint, and my subsequent tests were carried out on the Newmarket road, always starting from a certain mile-post. By this date, 1904, the wick or surface carburettor had been almost completely superseded by the liquid spray type and I had fitted a French spray carburettor known as the Longuemare, similar to that fitted to my grandfather's De Dion car. I chose it because I knew its habits well and how to adjust it. In this carburettor the air intake was near the bottom, and I had found that a good deal of petrol was being sprayed out from the air intake due, no doubt, to rebound when the inlet valve closed, and this was probably made worse by the fluttering of that valve. To catch and utilise this lost fuel, I fitted what amounted to a rudimentary wick carburettor to the air intake consisting of a cocoa-tin in which I placed a coil of lamp wick to absorb the spray. Air to the engine entered through holes punched into the lid of the tin, passed over the wick thus vaporising the petrol it had absorbed, and thence through the spray carburettor to the cylinder. With the economiser fitted I found that I could reduce the size of my carburettor jets and my petrol consumption by something like 20 per cent.

As a next step I raised the compression ratio by fitting a fairly thick shim between the inner and outer members of my two-part piston. This caused my engine to knock alarmingly at full throttle and optimum ignition advance, but not at the small throttle opening I needed to use during the competition, and in any case on the few occasions I should need full throttle I could always get rid of the knock by retarding the ignition. This increase in compression ratio gave me a further very substantial reduction in fuel consumption under my intended cruising conditions.

In order to have a fine control over the fuel/air ratio, I fitted an air bleed between the carburettor throttle and the inlet valve of the engine controlled by a screw-down needle valve.

Lastly, I had read in some technical journal an article on lubrication and friction in which it was pointed out how large a part viscous friction could play in certain applications such as the high-speed spindles used in cotton-mills, and in domestic sewing-machines. The article went on to point out that in cotton-mills a large saving in power consumption was effected by the use of sperm oil which had a very low viscosity, but a high coefficient of oiliness. At that time the favoured oil for petrol engines was that known as Price's Heavy Gas Engine Oil, a very viscous lubricant, but after reading this article, it struck me that in my engine with its internal flywheels swamped with oil, viscous friction might be playing a very significant part. I therefore decided to try using sperm oil. On enquiring I learned that what was sold as sewing-machine oil was in fact sperm oil, and I found an ironmonger who stocked it in small quantities. At that time it was the popular belief that to be effective a lubricant must have plenty of 'body', but sperm oil appeared to have none. It seemed almost as limpid as paraffin, and I was nervous about trying it but comforted myself by the thought that during the competition my engine would be only very lightly loaded. I therefore took my courage in both hands and tried it out. On my short test run it increased my mileage by at least 10 per cent, so I decided to use it in the competition.

On the afternoon of the competition the weather was perfect and the roads dry. There were, if I remember right, about fifteen competitors with a variety of engines of cylinder capacities, apart from mine, ranging from 150 to about 600 c.c., and most of them, like my own, had automatic inlet valves and spray carburettors. Vague rumours had got around that I had been playing tricks with my machine. From the general conversation at the starting-point I gathered that the consensus of opinion was in favour of going as fast as you could on the general grounds that the faster you went the further you got.

After having our tanks thoroughly drained and replenished with one quart of petrol and any pedal chains removed, we set off one at a time. Most of the competitors were off to a fine burst of speed, and were soon lost in the distance in a trail of dust, but it took me the best part of the first mile to arrive at the optimum conditions of throttle opening, mixture strength and ignition timing to give me a steady level speed of about 25 m.p.h., and I hoped that I would meet with no obstructions which would force me to disturb my nicely balanced adjustments.

Before I had reached Royston I had been overtaken by the later starters who waved to me as they passed. At a few miles after Royston I passed the first victim sitting with his machine by the roadside. He had covered only about sixteen or seventeen miles; thereafter at intervals of a mile or so I passed most of the rest of the team. By the thirtieth mile I had passed all but two or three, and was still going strong. These I overtook during the next three or four miles and finally came to a standstill only a little short of my fortieth mile, and several miles ahead of my nearest rival riding one of the smallest engines in the field.

That win, although in a very minor competition, stands out in my memory as my greatest triumph, for I had put my whole heart and all my thoughts into its achievement. I had, of course, every advantage in my favour, for it had been my good fortune to have been on intimate terms with the petrol engine since its first appearance in England, while in the hands of most of my competitors the motor-cycle was but a newly acquired toy. Also the conditions on the day of the competition were ideal in that throughout the whole run the level road was almost completely free of traffic, and I was able to maintain my optimum cruising conditions throughout practically the whole run.

Hopkinson, who had been watching the competition closely, congratulated me and called upon me to explain exactly what I had done and my reasons for doing it. I recounted all the modifications I had made to my engine and carburettor, and he cross-examined me in great detail as to the effects of each change, and to how nearly it complied with my estimate of its value. In particular he was interested in the use of sperm oil as a lubricant. I told him of the article I had read and, since my enclosed flywheels were awash, it had seemed to me that viscous friction might play an important part. This, he said, was a new line of thought to him.

A few days later he invited me to lunch and said that he had a proposition to put to me. He then told me that it was his intention to carry out a programme of research into the internal combustion engine with a view to satisfying himself as to the factors limiting the performance of such an engine, and that he would like to have my assistance. He went on to say that he had discussed the matter with Dykes and Inglis who agreed that I had the makings of a good mechanical engineer but took a dim view of my prospects as a civil engineer. If I accepted his invitation it would mean giving up all hope of taking an honours degree, but that hope was faint in any case. On the other hand, all were agreed that I would have no difficulty whatever in taking a pass degree. He recalled what I had told him previously, that I was destined to become a member of my grandfather's firm, but felt that a good mechanical engineer would be of far more value to the firm than a second-rate civil engineer. He pointed out that civil engineers were becoming more and more dependent on mechanical engineers for the tools of their trade, which had already changed from the spade and the wheel-barrow to the mechanical excavator; from the sledge-hammer to the pneumatic or hydraulic rivetter, and that in the years to come the civil engineer would become ever more dependent on mechanisation. He went on to say that in his view the internal combustion engine would sweep the board in all forms of transport on the roads and in the air if and when it could be produced in a light enough form.

I was, of course, immensely flattered by this invitation, for I could imagine nothing I would like better than to carry out research under his direction. He told me to discuss the matter with my parents and said he would gladly see my father and explain the position to him. Straightaway I wrote a long letter to my father telling him all that had passed and saying how much I would like to accept Hopkinson's invitation. This brought my father up to Cambridge a few days later when he had an interview with Hopkinson at which I was not present. He came out from this interview as much impressed as I had been and was all in favour of my accepting Hopkinson's invitation. I was now in the seventh heaven and cared little how I performed in my end of term examination. I did in fact make a very poor showing.

During the few weeks that remained of that term I saw a good deal of Hopkinson who told me that by the autumn he hoped to have completed his reorganisation of the Mechanical Sciences Department and would have more time to devote to research. He said he had been much impressed by what I had told him of my experience with pencil indicators as applied to high-speed steam engines at Willans' works and of the troubles we had experienced due to inertia of the moving parts, to friction and to backlash in the several joints of the indicator mechanism, all of which grew worse as the engine speed increased. He told me that he had designed an optical indicator which he hoped would be free from these troubles and that a prototype of it was being made in the Instrumentation Department and that one of my first jobs would be to try to make it work properly. He told me also that the firm of Crossleys had presented the laboratory with a new and up-to-date single-cylinder 40 h.p. gas engine and that he had earmarked this for our research.

Hopkinson's methods were by no means always orthodox; he believed in following up, step by step, a logical and reasoned sequence but only to a point: if that looked like becoming too prolonged then he would fall back on the principle of trying every bottle on the shelf, and if that did not achieve his end his next step was to try something really silly and see what happened. He taught me never to accept anything at secondhand unless it accorded with one's own commonsense and experience; to be sceptical of one's own observations when they failed in this respect, and never to cling too long to a theory, however cherished.

Hopkinson's design for the indicator was most ingenious and the very first version we made in the workshop worked beautifully after very little detail development. With the help of this indicator we started out to investigate the performance of the gas engine and, in parallel, we carried out combustion experiments in closed vessels, as others had done before us. At once he drew my attention to the fact that the rate of burning of the same combustible mixture at the same temperature and pressure was enormously more rapid in the engine cylinder than in the explosion vessel. He suggested that this might be due to the spreading of flame by turbulence in the engine cylinder, and proposed that we should put inside the explosion vessel a small electric fan. This done we found that the rate of burning could be speeded up enormously and that nearly in proportion to the speed of the fan. Sir Dugald Clerk, working quite independently, had demonstrated the same thing but he reversed the process, for he motored an engine with all valves closed and ignition cut off in order to allow the turbulence to subside; he then switched on the ignition when the rate of burning was found to be so slow as to be quite out of court even for a very slow-speed engine.

These two independent observations were of first-rate importance, for they showed, for the first time, how vital a part turbulence plays in the functioning of an internal combustion engine.

Hopkinson had his eyes fixed always on high speeds and prophesied that, given sufficient turbulence, there was no limit in sight to the speed at which an internal combustion engine could run, and run efficiently; and even today no such limit is in sight.

We concentrated all our attention on exploring the limitations of that gas engine. We studded it with thermo-couples and measured the temperature gradients through the cylinder walls and piston; we raised its compression ratio till pre-ignition set in; we then water-cooled the exhaust valve and raised it yet again, all with the object of finding out what factors set a limit to the power and efficiency obtainable. At that time, and wih Cambridge gas, pre-ignition (that is to say, self-ignition in advance of the timed spark) proved to be one of the limiting factors, and Hopkinson set out to investigate this. For this purpose, he inserted into the cylinder through the breech end a steel plug fitted with a thermocouple at its inner end. The further the plug projected into the cylinder the hotter it became until eventually it reached a temperature which ignited the working fluid before the passage of the spark. By projecting the plug further into the cylinder we could induce pre-ignition at any point in the compression stroke, and, at the same time, could observe the temperature at which it was initiated. We found of course that, as premature ignition set in, the power fell off, accompanied by a dull thumping, due to reversal of load on the connecting-rod bearings, but as the ignition grew earlier, the thumping ceased and eventually the engine, running quite smoothly, came gently to a standstill. Our indicator diagrams showed no change in rate of pressure rise, nor did the conditions differ in any way from those which obtained when we over-advanced the ignition. None of these findings were in the least surprising – I mention them only in view of what I am going to say about the petrol engine.

The petrol engine of that date, and indeed of today for that matter, had a habit of knocking sharply under certain conditions. The knocking was universally attributed to premature ignition, brought about by some overheated surface inside the cylinder. The knock in the petrol engine was, however, a high-pitched ringing noise, quite unlike the dull thud produced by premature ignition in the gas engine. During 1906, my last year at Cambridge, we managed to borrow a four-cylinder, high-speed Daimler petrol engine of the most up-to-date design, and this also Hopkinson put through an equally critical examination. We applied his optical indicator and found that we could get really excellent indicator diagrams at speeds up to 1200 r.p.m. Under certain conditions this engine would knock in the most alarming manner. We tried hard to secure indicator diagrams under knocking conditions but without success, for every time a knock occurred in the cylinder we were indicating, the mirror of the indicator was either shot out of its frame or shattered, a thing which had never occurred when indicating the gas engine. Hopkinson was very puzzled and intrigued by this and, with his usual quick perception, he attributed it to the setting up of an explosion wave inside the combustion chamber, a wave whose impact was sufficient both to shatter the indicator mirror and to set the cylinder walls in vibration, thus causing the high-pitched 'ping'. Since the same phenomenon could not be reproduced in the gas engine, we suggested that this 'detonation' as he called it must be a characteristic of the fuel. He would have liked to have investigated the question further but the engine had to be returned to its owner and there, for the time being, the matter ended. Our research occupied the whole of my second and third years as an undergraduate and for one term as a post-graduate student, for I had succeeded in taking a pass degree in Mechanical Engineering. Throughout all this period I was in almost daily contact with Hopkinson, and he treated me rather as a partner than as an assistant. As time went on my admiration increased for his genius in spotting the essentials of any problem and devising simple and convincing tests to provide the answer we were looking for. He was never depressed by our many failures, nor impatient over the delays and mishaps inseperable from this kind of research. He would propound a theory and suggest a test to prove or disprove it : if the former he would be delighted, if the latter he would roar with laughter and say 'now it is up to us to propound another and better theory'.

The primary objective of our research had been to break down and define what really were the obstacles confronting us in the design and development of a really light and efficient petrol engine for aircraft. By the end of my career at Cambridge Hopkinson had concluded that we had carried our research as far as we could go, for the time being at all events, and that it remained for the mechanical designer to take the next step. He told me that the Admiralty were pressing him to pursue certain other lines of research unconnected with internal combustion engines and that he hoped I would carry on, in particular, he said, into the phenomenon of knock in the petrol engine. I told him I had every intention of carrying on, at all events in my spare time.

As a parting present he gave me his optical indicator which I was to find invaluable in the years that followed. I was very sad at parting from Hopkinson whom I both liked and admired so much. To this day I can recall the sound of his boisterous laugh as he strode down the passage leading to our laboratory. It is a thousand pities, I think, that Hopkinson published so little of the work we did together, but he always had a strange reluctance to put pen to paper. He left it to me to reap the harvest he had sown and to be given much of the credit which was rightly due to him.

I do not wish to give the impression that while working with Hopkinson all my time was spent either in the workshop or the laboratory. Hopkinson was usually fully occupied in the mornings with lectures and administrative work and left me to my own devices to carry on such experiments as he suggested, or to prepare and rig up such equipment as I thought necessary for our joint work in the afternoons. In this I enjoyed the assistance of a very intelligent young lab. boy on whose resource and initiative I could rely. Since most of our joint research was carried out during the afternoons, I could not take any active part in sports involving team work, such as football or rowing, but I had plenty of time for such games as squash rackets in winter and tennis in the summer terms, and to a rather childish but exhilarating sport which Thornycroft, Sassoon and I had invented. On the eastern side of the road to Ely, some three or four miles out of Cambridge, lay a very large area of waste land, thickly overgrown with low scrub such as brambles and gorse, and intersected with various narrow footpaths. The whole of this area simply swarmed with rabbits, but except in the blackberry season it was otherwise quite deserted. Mounted on our motor-cycles, armed with small .22 revolvers and keeping a safe distance apart, we would charge down these narrow and very bumpy footpaths as fast as we dared, blazing away with our revolvers at the rabbits as they scurried across the paths ahead of us. Needless to say we had many spills but it was soft falling into the brambles or scrub on either side of the paths, and neither we nor our machines suffered any serious damage, while as to the rabbits my conscience is clear that not one of them suffered the slightest physical injury at our hands.

I have nostalgic memories, too, of the many happy winter evenings when, with a group of friends, we sat smoking our favourite pipes in front of a roaring fire in my large sitting-room in Trinity Great Court when we discussed how to put the world aright with all the assurance born of inexperience and over-simplification.

CHAPTER 7

The Dolphin Venture

Before starting work with Hopkinson I had become rather intrigued by the possibilities of the two-cycle engine using a separate pumping cylinder as described by Dugald Clerk in his classic book on *The Gas and Oil Engine*. At that time there were on the market a number of small two-cycle petrol engines with crankcase compression and no valves. They had the virtues of cheapness and simplicity but they had many vices, not the least of which was their erratic behaviour at reduced loads, and inability to run idle without misfiring or stalling. Mindful of my experiences with my pumping engine I hoped that in a new design by employing end-to-end scavenging of the cylinder, and by the use of a bulb in the cylinder head through which the ingoing mixture from the pumping cylinder would pass, I would be able to obtain a measure of stratification sufficient to give me good idling. To achieve this I would have to use an automatic inlet valve which on light loads would not open until near the end of the pumping piston stroke; thus at light loads only the bulb would contain combustible mixture, while the lower part of the cylinder would be left full of residual exhaust products. Hence, on light loads the conditions would be similar to those obtaining in my pumping engine excepting only that the main content of the cylinder would be residual exhaust products instead of air. On full load the pumping cylinder would, of course, start delivering much earlier in the scavenging period and the charge would then pass through the bulb into the main body of the cylinder, driving out the exhaust products through ports uncovered by the piston at the bottom of its stroke, thus leaving the cylinder full of combustible mixture. I realised, of course, that in the size of the communicating passage between bulb and cylinder I should have to compromise. It would have to be large enough for the whole scavenging charge to pass through on full load, yet small enough to provide adequate segregation

96

between bulb and cylinder when idling or at light load.

I planned to arrange the cylinders in vee form with the pumping piston linked to the big ends of the working connecting-rods. Here again I had to compromise. From the point of view of dynamic balance the angle between the cylinders should be 90 degrees; from that of scavenging a closer angle would be preferable. I compromised with an angle of 75 degrees. I realised that the scheme depended on the use of my *bête noire,* the automatic valve, but there seemed no way of avoiding this. I discussed my scheme with Dykes and many other of my friends, including Michael Sassoon, first cousin of Oliver Thornycroft, who was enthusiastic about it. Michael Sassoon was more than a year senior to me and a very skilled mechanic; he at once offered his help in the building of such an engine.

Sassoon at that time was a gentleman of leisure. His father, a wealthy banker and a brother-in-law of Sir Hamo Thornycroft the sculptor, had recently died leaving Michael a man of means. His one interest in life was workshop practice and he spent nearly the whole of his time in the university workshops, and between times in his own well-equipped workshop at home. Another of my friends, Harry Hetherington, was also enthusiastic and helped me to prepare such drawings as were necessary to enable Sassoon to start work on an experimental engine. In the event, Sassoon did all the pattern-making and most of the machining for I was just starting work with Hopkinson and was kept busy making up pieces of equipment for our research. Sassoon and Hetherington, more ambitious than I, agreed to start out with a two-cylinder version of a size suitable for a light car. With the able help of several other friends, including Hamilton Gordon (later to found the firm of Weyburn Engineering) the engine was completed in a remarkably short time. I had, of course, told Hopkinson all about it. He was encouraging and though he thought I was rather over-optimistic said that, if it turned out as I hoped, it would make a very useful engine. After a good deal of fiddling about we got it running quite well. It certainly idled beautifully and would tick over at a very low speed, but its maximum power output was rather disappointing due, as we found later, to inadequate breathing capacity, high pumping losses and high mechanical friction.

Our scavenging charge on its way from the carburettor to the working cylinder had to pass not only through two spring-loaded automatic valves in series but also through a long transfer pipe connecting the pump and working cylinders, and finally through the restricted passage between the bulb and the cylinder, all of which cost us dear both in negative work and in volumetric efficiency.

The cylinder bore of this first prototype was $3\frac{3}{4}$ inches and the stroke 4 inches, a total cylinder capacity of 1.44 litres. To the best of my recollection the maximum power output was about 10 h.p. at any speed between 800 r.p.m. and 1200 r.p.m. On the credit side the engine was smooth running, relatively quiet and flexible. Sassoon and Hetherington were delighted with its performance and Hopkinson and Dykes were encouraging and said they thought we had done very well at our first attempt and both made valuable suggestions as to how it could be improved upon. Cavendish Butler, another of my friends, an Irishman whose home was on an island in Lough Erne, thought it would make an ideal marine engine, in which service its limited speed range would be no handicap. Incidentally, he later acquired this actual engine, installed it in a hull of his own design and making, and used it regularly for several years.

In the summer term of 1905 my cousin Ralph with whom I had almost completely lost touch since he left school turned up unexpectedly at Cambridge, having just completed his three-year apprenticeship at the Arrol-Johnston works in Scotland. I introduced him to Sassoon who showed him our engine on test. He was much impressed by its performance, and before I knew what was happening he and Sassoon had made plans to form a company for the manufacture of the engine and complete motor cars. Ralph, as I have said, was an incurable optimist, a go-getter and one who did not let the grass grow under his feet. While in Scotland he had made many friends, among them an apprentice draughtsman named Thornton, a year or two older than myself, whom he wished to bring into the partnership. My part, apparently, was to be that of honorary consulting engineer and designer. Ralph, as usual, foresaw the glittering fortune awaiting the venture. Sassoon saw an opportunity for his skill as a production engineer but was not much interested in the commercial side. Thornton was to be in charge of the Drawing Office and General Secretary, while Ralph himself was to be Managing Director and Salesman.

In the course of my school holidays, Ralph and I had come to know the area of Shoreham-by-Sea, its harbour and the River Adur on which we used to sail. We agreed that this might be a suitable area in which to start operations, so he dashed off to Shoreham and soon located a small but derelict shipyard within the precincts of the harbour. It comprised a fairly large shed to serve as machine and fitting shop, a small office, a jetty to which boats could be moored, and a slipway. This he earmarked as an ideal site for the start of the venture.

There remained only the small matter of finding the necessary capital. Ralph had no money beyond a small allowance from his father in Australia, and I was in the same position. Thornton had inherited a little money but the bulk of the capital was, I believe, provided by Sassoon, though my uncle, Herbert Rendel, contributed a share but took no active part in the venture. My father smiled on this project and remarked characteristically 'you young people will gain experience and have a lot of fun, but don't expect to grow rich on it'. He suggested christening the engine 'The Dolphin' and he modelled a beautiful little bronze dolphin to be mounted on the radiator of the car.

While our first experimental engine was on test, Hetherington and I set about designing an improved version in which we embodied all that I was learning from my research work with Hopkinson. We decided to increase the bore of the working cylinders to 4 inches and that of the pumping cylinders in like manner, to provide for larger valve area and to reduce the mechanical friction by mounting the crankshaft on ball bearings.

Hetherington, like Sassoon, was a gentleman of leisure. His parents were the owners of a large country estate known as Berechurch Hall, near Colchester which he was some day to inherit and to equip him for this he was reading for a Degree in Law, but his hobby, like mine, was that of engine design. Though he had had no engineering training, he was a mine of information on the state of the art in those early days, and had taught himself to be a first-rate mechanical draughtsman. We spent many happy evenings together in his rooms or mine scheming out designs for engines. In this we were sometimes joined by Oliver Thornycroft and Michael Sassoon's younger brother, Hamo, also a great enthusiast.

I must now go back a couple of years. During my rambles in London I had come across, among others, the firm of Lloyd & Plaister, who owned a small but magnificently equipped machine and fitting shop, and employed about twenty highly skilled mechanics who specialised in the making of experimental or prototype machines and equipment of all kinds. Lloyd was a man after my own heart. He was, I believe, a wealthy man in his own right and he lived with his mother in a beautiful house and garden close to Alexandra Palace, and I imagine ran his establishment in the Finchley Road more as a hobby than as a commercial enterprise. Like my friend Hamilton Gordon, he loved tackling difficult jobs in his sphere of mechanical engineering and was himself an expert mechanic, regarded with both respect and affection by his employees. His partner, Plaister, looked after and relieved him of all the correspondence and business side. Lloyd and I made friends at once, for we had so much in common.

When in 1905 I told him about Ralph's project to manufacture our two-stroke engines at Shoreham, he at once offered his help and advice, both of which proved most valuable.

It took several months of exasperating legal discussion before Ralph was able to get possession of the site at Shoreham and have the Company registered under the title of 'The Two-Stroke Engine Company', and it was not until early in 1906 that the three partners, Ralph, Sassoon and Thornton, were able to start preparing and equipping the premises. In the meantime, it had been agreed that two engines to Hetherington's and my improved design should be built, one to supply power to the machine shop and the other for tests and demonstration purposes. Both engines were built by Lloyd & Plaister and, when ready were tested at their works. Compared with our original experimental engine, they gave a considerably improved performance, with a maximum power output of between 15 h.p. and 16 h.p. over the speed

range of 1000 r.p.m. to 1500 r.p.m. In the meantime, Ralph and
Sassoon had been scouring the neighbourhood in search of second-
hand but serviceable machine tools. So far as I can remember
they acquired one huge and antiquated planing machine, three
lathes, two drills and a small shaping machine. Thornton, meanwhile, set up his drawing office and engaged a capable lad of
sixteen as an assistant draughtsman. As the company's Treasurer
and Accountant, he also acquired a cash-box and an account book.

It had been Ralph's ambition to build motor-cars. At that date
the British car industry was still in its infancy. Only a few leading
firms such as Arrol-Johnston, Napier, British Daimler, Wolseley
and Lanchester were just coming into production with cars of
their own design and, as yet, only on a small scale, for, of the few
cars in this country at that date, most were of foreign manufacture.
On the other hand, small garages were springing up all over the
country supplying petrol and oil and carrying out minor repairs.
It was the custom in those days, if you were a connoisseur, to have
your car tailor-made to your own taste and specification by your
local garage, who assembled it from the component parts supplied
by specialist manufacturers.

My uncle, Herbert Rendel, though not an engineer, had become an enthusiastic motorist and had ideas of his own as to what
he wanted. On my recommendation he had, at some time in 1905,
commissioned Messrs. Lloyd & Plaister to build him a rather large
car to his specification. The firm had already supplied two or three
cars to local clients and had specialised in the manufacture of
gear-boxes and clutches. All this was just before Ralph had appeared on the scene. My uncle had specified that the car was
to be propelled by a four-cylinder M.M.C. engine, but Ralph
succeeded in persuading him to hold his hand and wait until the
projected Two-Stroke Engine Company could supply him with
a Dolphin engine to be built to my design. In the summer of 1905
Ralph with his usual optimism had expected that all formal
matters concerning both the formation of the Two-Stroke Engine
Company and the acquisition of our site at Shoreham would be
completed in two or three weeks, and that by the end of that year
we should be under way with the manufacture of Dolphin engines,
by which time Lloyd & Plaister expected to have completed the

rest of the chassis. In the event we did not get possession of the site at Shoreham until the beginning of 1906, and were not in a position to start any kind of manufacture until well on into the spring of that year. Reluctantly, therefore, it was agreed that the engine for my Uncle Herbert's Dolphin car should be made by Messrs. Lloyd & Plaister. We had all been rather appalled by the size and weight of the chassis being built to my uncle's specification, and still more so by that of the luxurious five-seater body being built for it by a firm of coach-builders at Croydon, and it was becoming increasingly evident that our two-cylinder Dolphin engine, developing at that time a peak of barely 16 h.p. would not be man enough. It was therefore decided to make a four-cylinder version which hopefully would develop slightly more power than the 30 h.p. M.M.C. engine originally specified.

When designing the original two-cylinder prototype, Hetherington and I had always in mind the possibility of a four-cylinder version being required at some later date and our design had therefore taken this into account. Ralph, in the meantime, had collected castings and other material for the building of a dozen or more two-cylinder engines so that Lloyd & Plaister were able to start straightaway on a four-cylinder Dolphin engine for my Uncle Herbert's car. This was completed in very quick time and the car was ready for its first road test in July, 1906. After the usual teething troubles, it performed on the whole very well indeed, so well in fact that both my grandfather, my uncle Arthur Rendel and his cousin, Felix Wedgwood, each placed orders in the following year for four-cylinder engined Dolphin cars to be built at our newly established works at Shoreham.

It had been our original intention both to build relatively small light cars propelled by two-cylinder Dolphin engines and, at the same time, to supply proprietary Dolphin engines for assembly by other firms and I had spent many happy evenings during the autumn term of 1905 and the spring and summer terms of the following year laying out scheme designs for a light 16 h.p. two-seater car to weigh not more than 12 cwt. and in which some at least of the component parts of the chassis could be produced on our limited tooling equipment at Shoreham, thus rendering us less dependent on outside sources of supply. The car we planned would be less than half the weight of my Uncle Herbert's car and have at least as good a performance and would, we hoped, be fully competitive with contemporary four-cylinder, four-cycle engined cars. In this car we planned to use a four-speed gear-box of which the third speed would give a direct drive for normal use, the top speed being an overdrive for use on the comparatively few main roads whose surfaces were well enough maintained to permit of speeds of 40 m.p.h. and over. The construction of this prototype car was the first job undertaken at Shoreham, but it took a very long time to assemble, mainly because of difficulties and delays in getting such component parts as we could not make ourselves, and partly because of the sudden appearance of a dynamic little Scotsman who brought with him decided ideas of his own. Angus, as we all called him, had been a charge-hand at the Arrol-Johnston works during my cousin's apprenticeship there. He was indeed a 'character'; a rather short, stockily built man in the early thirties who, though almost totally uneducated had, we gathered, already seen a lot of life. He had acquired an almost fanatical devotion to my cousin Ralph, whom he admired above all other men. Angus's hobby was fishing. In his spare time he used to go out with the Clyde herring fishery fleet and so had become very knowledgeable about the technique and handling of drift-nets and trawls. On seeing our site at Shoreham with its small jetty and slipway, he at once urged that we should turn our attention to the engining of the local fishing-fleet at Shoreham, for which he insisted we were ideally situated, for not only had we the facilities for installing engines in fishing and other boats, but from all accounts our Dolphin engine gave just the performance required for the handling of drift-nets, particularly because of its ability to run dead slow for hours on end. To each of us this was a new line of thought, and it was reinforced by reports from my friend, Cavendish Butler, in Ireland that our original experimental Dolphin engine was giving a very good performance in his boat. At that time all the vessels of the Shoreham fishing-fleet depended on sails alone; indeed, apart from occasional visits by coastal steamers, there were no mechanically propelled vessels of any kind in the neighbourhood. The entrance to the harbour in those days took the form of a long narrow passage bounded on either side by high piers, and through this passage the large volume of water contained by the harbour and the estuary of the River Adur flowed in or out like a mill race, at certain stages of the tide reaching a velocity of 5 knots. Against such a rapid stream, no sailing-boat with sails blanketed by the high sides of the piers could possibly contend. Hence, it was only at certain stages of the tide that a sailing-boat could enter or leave the harbour and this, of course, was a heavy handicap to the fishing-fleet for if they missed the tide, they missed the market!

We had all been so obsessed by the glamour of the motor-car that we had thought little or nothing about other uses for our Dolphin engine until Angus opened our eyes to the yawning local market that awaited us on our doorstep. Angus, a good mixer, set himself to fraternise with the local fishermen, at first sight a rather forbidding lot when ashore but, as we later found, kindly and good-natured and scrupulously honest in their dealings with us. Angus was thoroughly at home in their company, for he could out-drink and out-swear the best of them, and had much to tell them about his fishing experiences in Scotland. The upshot of all this was that he managed to procure the loan of one of their vessels in which to install one of our two-cylinder Dolphin engines. From a firm in Southampton we procured a reverse gear and a suitable propellor.

Almost all the Shoreham fishing-boats were of similar size and design, between 30 feet and 35 feet long and 8 feet or 9 feet beam, each with a normal crew of six or seven needed to handle the huge area of net they carried. Hauled up on our slipway it did not take long to install the engine and gear. The first trial run was a great adventure. The speed through the water was just under 6 knots or sufficient to stem the stream through the harbour-mouth under the worst conditions. At the next spring tides Angus gave a demonstration of entering and leaving the harbour at all stages of the tide. This was a great success and he was asked to repeat it again and again with different groups of fishermen on board. Next, Angus and a party of fishermen went out for a night's fishing with the rest of the fleet. They reached the fishing-ground, about three miles off shore, in about half the time taken by the sailing craft. They reported that the engine would run slow enough and steadily enough just to keep the several hundred feet of drift-net out-stretched, the engine speed being only about 160 r.p.m. This satisfied the fishermen and delighted Angus who said that the Kelvin oil engines, then just coming into use by the Scotch fishermen, would not run slow enough for drift-net fishing. The immediate result of these demonstrations was that we were inundated by requests from the fishermen to install Dolphin engines in their boats.

For the next twelve or eighteen months our works at Shoreham was fully occupied in this work. For this our circumstances were ideal, for not only could we carry out the entire installation ourselves, but with our jetty and slipway we had all the facilities for servicing engines requiring overhaul or repairs, and the fishermen, for the most part, paid promptly, and in gleaming golden sovereigns. To see a whole heap of these spread out on the table gave an impression of wealth and prosperity such as no mere cheque, however large, could convey. As I recollect we charged £50 for an engine and just under £100 for a complete installation.

Whether or not we made any profit is anybody's guess, for none of the partners knew how to keep accounts, nor had they any understanding of the term 'overheads'. To test our financial position we rattled the cashbox, and that I suspect was about the only clue we had. If we did not make a profit we had only ourselves to blame, for while the boom in marine engines lasted, we had a golden opportunity.

By the end of 1907 our local market for marine engines had become saturated and in spite of efforts by Angus and Ralph, all our attempts to sell marine engines in other areas came to nothing, for by that time the larger firms such as Thornycrofts, Kelvins, Parsons and others had spread their tentacles into most of the south-coast ports and could offer better prices and far better servicing than ourselves. They could supply spare parts which would fit; ours would not, for each of our engines was tailor-made, and though quite as well made as those of our competitors, their individual parts were not interchangeable. This did not matter so long as we could carry out the servicing and any renewal of parts on our own premises, but formed a fatal barrier to the use of our engines elsewhere.

With the bottom fallen out of our marine engine market we returned once again to our original programme, that of producing Dolphin motor-cars and our first step was to complete our prototype 16 h.p. car which had been left lying on the shelf while the marine engine boom lasted.

During all that period I personally had been almost completely out of touch with Shoreham, partly because I was cramming for the Associate Membership Examination for the Institution of Civil Engineers, and partly also because while the boom was on there was little for me to do. By the time Hetherington left Cambridge in the summer of 1906, he and I had completed our scheme designs for both the two-cylinder and four-cylinder cars and had handed them over to Thornton to elaborate and detail them. By the end of 1907, with my much dreaded examination behind me, and my position in my grandfather's firm established, I was able, once again, to take a more active part in the affairs of the Two-Stroke Engine Company, to pay visits to the works and to carry out tests on the engines with a view to improving their performance. For this purpose Hopkinson's optical indicator proved invaluable; with it I was able to take reliable light spring indicator diagrams at speeds up to 1200 r.p.m. These revealed both that the volumetric efficiency of my pump cylinders was lower, and the negative pumping work higher than I had thought and that the poppet inlet valve was the chief offender.

Some years earlier while prowling round Devonport Dockyard I had noted that during refits, a new form of thin sheet metal inlet valve was being fitted as a replacement of the old-fashioned leather flap valves in condenser air pumps. An engineer-artificer had told me that these reed valves were made of a special copper alloy known to him as Kinghorn metal, which was particularly resistant both to fatigue and corrosion, and that they imposed much less restriction to the flow of air than did those which they replaced. This observation, like so many others, had lain dormant in my sub-conscious mind, but was awakened when I examined my indicator diagrams. From an advertisement I discovered that the material was supplied by the Metallic Valve Co. of Liverpool, from whom I obtained some sample sheets of varying thicknesses, together with some useful information. We therefore redesigned the pump cylinder heads to incorporate this type of valve in the form of a number of radial fingers, each covering a row of drilled ports. After some experiment these valves proved far more reliable than the spring-loaded poppet valves we had used previously, and they gave us a better volumetric efficiency and a marked reduction in negative pumping work, with the result that we were able to step up the maximum power of our two-cylinder engine from 16 h.p. to between 19 h.p. and 20 h.p., and its effective speed from 1500 r.p.m. to nearly 1800 r.p.m., and that with a marked reduction in engine noise. This improvement rendered the engine much more attractive for automobile use.

As I have mentioned earlier, the success of my Uncle Herbert's first Dolphin car brought us orders from three of my Rendel relatives for four-cylinder cars or at least chassis, for the bodies in each case were ordered from coach-builders. Ralph had also secured orders for two more, one from a Brighton jeweller and another from a local butcher, but nobody seemed to want our smaller and much lighter two-cylinder version which, when completed and fitted with Kinghorn valves, gave really a fine performance, with a top speed of just over 50 m.p.h. on the overdrive, and a very smooth, quiet and comfortable cruising speed of about 25 m.p.h. on the direct drive. Since all our firm orders were for the larger four-cylinder version, it was decided not to proceed any further with the smaller one, which I acquired for my own use and which served me very well for the next ten years.

In all we constructed or assembled I think eight four-cylinder chassis of which I think Sassoon had one, a friend of Hetherington's another, and the eighth kept as a works car. All, so far as I know, behaved quite well, especially my grandfather's in which he went every year for a long tour in the West Country. The work of assembling these last few cars proved a tedious and exasperating business for the makers of component parts were constantly changing their designs to suit the tastes of the larger car manufacturers, who placed bulk orders with them, and we had to make do as best we could with what was available. By this time also the three partners, even including Ralph, were becoming thoroughly disillusioned as to the prospect of a company such as ours ever being able to compete with the large firms with their country-wide sales and servicing organisations and large capital resources. In addition the multi-cylinder, four-cycle engine was improving by leaps and bounds, while we saw little prospect of much improvement in the performance of our two-cycle Dolphin engine.

Other factors also entered into the picture: Ralph had married and must now look for a more assured source of livelihood; Sassoon had become engaged to a Canadian girl and was wanting to live in Canada, and Thornton, too, was getting thoroughly despondent. It was decided, therefore, by common consent, to wind up the Two-Stroke Engine Company as soon as the batch of cars on order was completed.

Thus, on rather a sad note, ended the reign of the Two-Stroke Engine Company after nearly four years of existence. So long as the shipping boom had lasted it had been a very happy time for all concerned, but the last year had been one of boredom and disillusionment.

To the best of my knowledge and belief, the three partners each drew only a token salary. Apart from that the only paid staff consisted of Angus, two young but skilled mechanics, and Thornton's junior draughtsman.

Ralph with his wife went to India to act as technical adviser to one of the Indian princes who was keen on mechanical engineering. There, I gather, he had a very good time and in India he remained until he retired some twenty-five years later. He never quite succeeded in catching up with the large fortune he was always in pursuit of, but such was his temperament that I think he enjoyed the pursuit more than he would have done the capture.

The two mechanics got jobs locally, while the junior draughtsman went to the Brighton Locomotive Works, where he ultimately became Chief Draughtsman. I do not know what became of Angus except that he returned to his native Scotland.

When some ten years later I returned to Shoreham the whole scene had changed completely. The then newly acquired week-end habit, the lure of the seaside, and the advent of the motor-car, had converted what had been a self-contained fishing village into a flourishing holiday resort: along the hitherto deserted sea-front to the west of the harbour entrance there had sprouted a ragged line of buildings ranging from ornate bungalows to plain lodging-houses and discarded railway-carriages. Gone was the busy fish-market; of the fishing-fleet. with its Dolphin engines no trace remained. I recalled that the engines had been designed for motor-cars, hence their crankcases were of aluminium, and their crank-shafts mounted on ball bearings, both of which might well have dissolved like sugar after a few seasons in salt-water, but in those earlier years we had not thought of this possibility. As to the fisher-men themselves, some had left the neighbourhood and those who remained had found that catering for visitors was less arduous and more remunerative than fishing.

I think the winding up of the Two-Stroke Engine Company marked the end of the Dolphin car as such, but not the end of the Dolphin engine. During the years 1906 to 1909 the firm of Lloyd & Plaister had prospered and expanded considerably. Lloyd himself had always taken a fatherly interest in our Two-Stroke Engine Company and had given us much valuable advice, some of which we had ignored. For example, he had warned us not to attempt to compete with the established firms in the production of large cars, but into this field we had been drawn by force of circumstances. He had warned us, too, that if we were going into production on any worth-while scale, we must ensure that all our working parts were strictly interchangeable. This advice also we had ignored.

On the winding-up of the Two-Stroke Engine Company he came to me with a new proposition which he said he had had in mind for some time, but had held his hand lest it take the wind out of our sails.

At that time there had grown up a wide demand for what were known as 'cycle cars'; that is to say, very light four-wheeled vehicles, usually two-seaters, which could be produced and operated very cheaply. Such vehicles were almost invariably fitted with single-cylinder engines of the De Dion type, and in such vehicles a single-cylinder engine, with its lack of dynamic balance and its large torque reaction, was an unpleasant companion. So long as the four-stroke cycle was adhered to, the only practicable alternative was that of the miniature four-cylinder engine, but such an engine in the then state of the art would be far too costly.

By contrast Lloyd maintained that a small two-cylinder Dolphin engine, with its absence of any valve mechanism, would be very much cheaper to produce, and would have all the advantage in the way of dynamic balance and of even turning moment that the four-cylinder, four-cycle engine could offer. His proposal, therefore, was that he would design such a car to weigh not more than 8 cwt., nearly all the component parts of which he could make himself, and he would ask Hetherington and myself to design for him a very light two-cylinder Dolphin engine, with a total cylinder capacity of about 700 c.c., and that he would pay us a royalty of thirty shillings on every car he sold. To this we readily agreed. He requested that the working cylinders should be vertical, and the cooling by thermo-syphon. The little engine we designed had a cylinder bore of $2\frac{3}{4}$ inches and a stroke of $3\frac{3}{4}$ inches. We provided for reed valves in the pump cylinders which were rather larger in proportion than in the larger Shoreham-built Dolphin engines, the accent being on power output rather than fuel economy. This little engine performed very well. Its maximum power output was about 12 h.p. at a speed of about 2000 r.p.m. and its maximum b.m.e.p. rather over 70 lb. per square inch. We had previously funked any attempt at using a reed valve of the Kinghorn type in the working cylinder on account of the high temperature and pressure. However, in this small engine it seemed possible that such a valve might survive, and in fact it did survive, and by its use we were able to increase the maximum power output to 14 h.p. and the maximum speed to about 2200 r.p.m.

Lloyd's first cycle-car was completed early in 1911 and it gave a good performance on the road. As we had expected, its good dynamic balance and even turning moment made it very pleasant to drive. Once satisfied with the behaviour of the engine, Lloyd set himself to prepare jigs and gauges to ensure interchangeability of all the working parts; meanwhile he carried out drastic routine tests on his first prototype car. Early in 1912 he put through a batch of, I think, twenty such cars to test the market. These sold readily and on them we received a royalty of £30, the first royalty we had ever received on a Dolphin engine. Thereafter, and until the outbreak of war in 1914, Lloyd & Plaister continued to make and sell Vox cars. In all they must have made well over 100 before the war put a stop to their activities. Plaister, who I believe had been a Naval reservist, was called up for active service, and Lloyd died, I believe, of a broken heart, for his establishment was taken over for the manufacture of munitions; thus ended the reign of the Vox car. Lloyd's death was a great blow to me, for he had long been a good friend and a wise counsellor and I missed him sadly.

While I was negotiating with Lloyd & Plaister after the winding up of the Two-Stroke Engine Company in 1909 Hetherington had got into touch with the Britannia Engineering Co. of Colchester, makers of general purpose lathes and other machine tools. Mr. Nicholson, their Managing Director, was anxious to extend the activities of their Company in other directions and he thought that our Dolphin engine would be admirably suited for dynamo driving. He therefore invited us to design two sizes of single-cylinder engines of $2\frac{1}{2}$ and 5 k.w. electrical output for the purpose.

Of the two sizes of engine we designed for the Britannia Co., the larger was of the conventional single-cylinder type. This was mounted on a common bedplate and direct-coupled to a generator. The bore of the working cylinder was $4\frac{1}{2}$ inches and the output of just 5 k.w. was at a governed speed of 800–900 r.p.m.

In the case of the smaller unit of $2\frac{1}{2}$ k.w., we departed from convention in that the front cover of the dynamo formed the crankcase of the engine, thus no bedplate was needed and there was no aligning problem. For these smaller units, Messrs. Laurence Scott & Co. of Norwich supplied the dynamos with extended armature shafts carrying an overhung crankpin and balance-weight at one end, a belt pulley and sprocket for a chain-driven magneto at the other or commutator end. This arrangement provided a neat and compact generating set and was much cheaper to manufacture; it proved very popular indeed and the Britannia Co. had made and sold a great number of these before the outbreak of the 1914 war put a stop to their activities.

I installed, at Graffham, one of these $2\frac{1}{2}$ k.w. sets in place of my old pumping engine in 1912; it served well for it provided all the light and water for the next twenty-five years and was still going strong when the public supply from the grid reached Graffham and permitted a much more lavish use of electricity.

From these Britannia-built Dolphin engines we derived a royalty of, as far as I can remember, four shillings per k.w. I have no record of the number sold but to the best of my recollection our royalties from this source totalled over £100 before 1914.

The outbreak of the First World War brought the career of the Dolphin engine to a sudden close. How much longer it might have survived is an open question. Neither Hetherington nor I could see any hope of further improving its performance, whereas that of the four-stroke engine was increasing by leaps and bounds, more especially during the war years. Wisely, I think, we made no attempt to re-instate it after the end of the war.

Sir Dugald Clerk
By permission of the Inst. Mech. Eng.

Professor Bertram Hopkinson, from the portrait by Arthur T. Nowell,
painted in 1911
By permission of Cambridge University

Laurence Pomeroy
By permission of the Inst. Mech. Eng.

Dr Frederick Lanchester
By permission of the Inst. Mech. Eng.

CHAPTER 8

Start of a Career

My grandfather, Alexander Rendel, had always hoped to keep his firm a Rendel family affair. As I explained earlier, of his five sons, the second, William, and the youngest, Harry, were both trained civil engineers and had been on his staff for some time. In 1894 he took William into partnership by which time Alexander was in his middle sixties, and a very tired man after nearly forty years of single-handed control. After inheriting the business in 1854, he had been able greatly to expand the world-wide practice and though probably neither so versatile nor so ingenious an engineer as his father, he was extremely far-sighted and successful. He appears to have taken little interest in the affairs of the Institution of Civil Engineers, but through his wisdom and integrity had gained the complete confidence of Whitehall. The higher ranks of civil servants, more especially those of the Colonial Office and the India Office relied on him absolutely.

In 1898 my Uncle William died. This must have been a severe blow to my grandfather who immediately took into partnership his younger son, Harry, and for some reason which I have never been able to fathom, also a certain Mr. Robertson. Robertson, a man in the early fifties, was an experienced civil engineer and an active member of the Institution of Civil Engineers; thus, technically, he was well-qualified, but he lacked any real stature. By virtue of his experience he was made senior to Harry, and the name of the firm was changed to Rendel & Robertson.

At the time of Robertson's appointment, it was agreed also that as I was the only member of the Rendel family in the next generation with any interest in engineering, and provided that I qualified, I should become a member of the firm and eventually a partner in it. I was only thirteen but my grandfather liked to feel that some member of his family would have an interest in the firm, even though I did not bear the Rendel name. Thus,

113

throughout my school and Cambridge days I accepted the fact that I was destined to join the firm of Rendel & Robertson as soon as my education was completed.

My Uncle Harry was my favourite uncle; he was a really brilliant engineer and a perfectly charming companion. During my childhood days I had seen little of him for he was often abroad, but when he returned to take up the partnership we often met in the summer and at week-ends when he and I went shooting together. He told me of his experiences in India and Africa, and I looked forward to the future when we would both be partners in the firm. As senior partner he would look after the civil engineering side, while I would take care of the mechanical developments. Then, in 1904, my Uncle Harry died from septic pneumonia after only two or three days' illness.

To all of us, and to my grandfather in particular, this was a terrible tragedy. My grandfather was in his middle seventies and I had the impression that he was not very happy in his partnership with Robertson, and had been content to rely more and more on my Uncle Harry. Within a matter of days he took into partnership a senior member of his staff, Mr. (later Sir Seymour), Tritton. Tritton had served my grandfather for many years, first, as his confidential clerk and secretary, and later in charge of the business side of the firm. I do not think he ever claimed to be an engineer, but he was an able and conscientious man whom everybody trusted, and was a great support to my grandfather in the unhappy years that followed Uncle Harry's death, especially when dealing with Government red-tape which usually exasperated my grandfather.

As a result of these unheavals my grandfather gradually left the executive work to his partners, and though his mind was as alert as ever, he retired from government committees because perhaps like myself at the same age, he was getting very deaf. Such was the state of the firm when I joined it in the summer of 1907.

My last term at Cambridge in the autumn of 1906, like that at Rugby, had been rather a sad one, for most of my friends, including Welsh, Hetherington, Thornycroft and others had left at the end of the summer term. As a post-graduate, I had to give up my nice rooms in Trinity Great Court and go once again into 'digs', and there I became vaguely ill. Worst of all I hated winding up my work with and parting from Hopkinson, and since my Uncle Harry's death I no longer looked forward with eagerness to joining my grandfather's firm. On top of all this came the bombshell that Robertson insisted that if, as agreed, I was eventually to be a technical partner, I must at least become a Corporate Member of the Institution of Civil Engineers, and this meant that I must pass the Associate Membership Examination of that Institution. Dr. Tudsbury, then Secretary of the Institution, recommended a tutor who specialised in coaching candidates for that particular examination, and he certainly found a very good man for after seven or eight weeks of sheer cramming, my tutor succeeded, much to my surprise, in pushing me successfully through the examination. So I became, on paper at least, a fully qualified civil engineer, entitled to the letters 'A.M.I.C.E.' after my name. Luckily nobody has ever entrusted me with the design of a bridge or a dam.

After a time my health and spirits improved until I was fit enough to start work. My first assignment with the firm was that of an inspector of equipment needed for bridge-building. The head of our Inspection Department was a certain Mr. Dobbie who took a fatherly care of me and who, knowing my interest in engines, gave me the job of inspecting and witnessing the tests of steam, gas and oil engines, air compressors and hydraulic mechanisms. It soon transpired that most of the British equipment we needed could be found in Manchester and Glasgow, each of which I visited frequently, generally for two or three days at a time. It was during these visits that, for the first time in my life, I came into contact with some of the leading lights in the field of mechanical engineering. One day, while witnessing a test of a small oil engine at Crossley's works in Manchester, a very tall middle-aged man, whose name I forget, asked me if I was the young man who had assisted Hopkinson in his researches at Cambridge on the engine which Crossleys had presented to the University. When I pleaded guilty he said that from all accounts we seemed to have mauled it about pretty thoroughly and he invited me to tea with him in his office. Over tea he asked me many questions about Hopkinson's experiments, and I told him briefly about our work on turbulence and heat losses, and Hopkinson's belief concerning pre-ignition and knock and much else besides. He appeared to be very interested and said that if I could give him notice of my next visit, he would collect some of his engineer friends to meet me in the evening for he was sure that they would be as interested as he was in what I had to tell about Hopkinson's work. On my next visit I found a message from him inviting me to dinner at the Midland Hotel and saying that he had invited also five or six of his friends. At the dinner that evening he introduced me to Mr. Windeler, Chief Engineer of the firm of Mirrlees, Bickerton & Day, the leading firm in this country manufacturing large industrial diesel engines; to Mr. Chorlton, later Sir Alan Chorlton of Mather & Platt, manufacturers of large gas engines, dynamos, etc.; to Mr. Onions of Hornsbys, makers of the Hornsby-Ackroyd oil engines; to Mr. Davidson of Browett-Lindley, manufacturers of high-speed steam engines and air compressors, and to Mr. Bickerton, the Chief Engineer of the National Gas Engine Co. To me it was a great experience to meet such a distinguished gathering. They were all charming and did their best to put me at ease, and after dinner we adjourned to a small reception-room where we talked shop far into the night. That evening with the technical heads of the most important firms engaged in the manufacture of industrial engines was a wonderful event for me. In the first place, it brought home to me how much I owed to Hopkinson's teaching, and how much apparently new ground we had explored. To the repeated question 'Why had not Hopkinson published his findings?' I could only answer that it was not due to his secretiveness but rather to modesty and his hatred of personal publicity. I was surprised to learn how little research or development work was being carried out by these large firms. At that date, 1908–9, industrial gas, oil and diesel engines were at the heyday of their popularity. I was surprised to learn also that the leading firms in this country were licensees of, and obtained their designs from, German or other Continental firms; thus Crossleys were licensees of Otto Deutz; Mirrlees of the M.A.N. Co. and Mather & Platt of Körting, and so on. All these foreign firms it seemed were carrying out a great deal of long-range research and development, financed largely by the royalties received from British and other companies. My friends all regretted that this should be so, for it both hurt their pride and limited their activities as engineers, but it was at that time the belief of the business side of management that it was cheaper to buy designs and 'know-how' from abroad. Chorlton, in particular, was very resentful about this short-sighted policy. The complaint about the lack of research and development in this country had awakened my youthful dream that in the years to come I might one day have an establishment of my own in which to carry out such work on behalf of industry.

My friend Harry Hetherington got married soon after leaving Cambridge and he and his wife came to live in London. Once again I was in close touch with him and we spent many happy evenings together designing imaginary engines. Inspired by Hopkinson's enthusiasm, it had become my ambition to make, or at least to design, an aero-engine version of my two-stroke Dolphin, but gradually and reluctantly we had to abandon this idea for we could see so hope of reducing its weight or its fuel consumption to an acceptable figure.

So long as I lived in London I could not myself carry on with the research that Hopkinson had initiated, but by attending lectures and listening to the discussions that followed I managed to keep myself fairly well-informed. To me, the most interesting lectures were those delivered by Dugald Clerk, F. W. Lanchester and Pomeroy, Chief Engineer of Vauxhalls. All three were models of lucidity, and their lectures were delivered in a language that all could understand. I longed to take part in the discussions that followed, but at first had not the courage to do so. I noted that Dugald Clerk's replies to his questioners were invariably kindly and courteous, even though the questions were irrelevant, and that he was particularly kind and encouraging to the younger and more timid members of his audience. Lanchester, on the other hand, was not so patient, and indeed was often sarcastic to those who dared to disagree with him. At long last I screwed up my courage to question Dugald Clerk about his advocacy of the use of the stratified charge in petrol engines. Timidly I spoke of my experience of this as applied to my pumping engine, and later to my Dolphin two-stroke, and to Hopkinson's conclusion that because of the narrow range of burning of petroleum fuels, it would be difficult, if not impossible, to cover the whole range of load by stratification alone. He replied with his usual courtesy that what I had said was very much to the point, and that he needed to think about it. At the end of the meeting he invited me to call upon him at his London office to tell him more about my Dolphin engine and of Hopkinson's investigations into the range of burning of volatile hydrocarbon fuel. He was, as I expected, perfectly charming and seemed really interested in what I had to say. He then asked me about my future career and my ambitions. I told him that the height of my ambition was to follow on with the research which Hopkinson had initiated, and to design a really light and efficient aeroplane engine. He wished me well and told me not to hesitate to ask him for any help or advice I needed. This was my first meeting with Dugald Clerk and, as with Hopkinson, I fell a victim to the charm of his personality. He told me that he himself had had to give up the research and development work so dear to his heart and to concentrate on Patent Office work, which was much more remunerative, but it did not mean that his interest in the internal combustion engine had waned. After this first meeting I paid several more visits to his office when we talked together not only about internal combustion engines but also about his and my own early experiences. I found this very flattering, and one day he invited me to lunch with him at the Athenaeum, where he introduced me to Sir Charles Parsons, Professor Dalby and to Mr. Campbell Swinton, all Fellows of the Royal Society.

As time went on it was borne in upon me that much of the equipment I was inspecting was really quite unsuitable for its purpose which, in general, was the erection of bridges in such countries as India. In those days it was the responsibility of the consulting engineer not only to design the bridge but also to direct and supervise its erection. All that the actual contractor had to do was to provide the steelwork to the design and under the inspection of the consutants and there his responsibility ended. The equipmen needed for the erection of the bridge was provided by (as was the labour), and became the property of, the railways in India, or, in the case of the State Railways, of the India Office.

In many, if not in most parts of India, the rivers almost dry out during the long spell of drought, but become raging torrents during the monsoon, with as much as 300 inches of rain in three months, leaving wide areas of flooding on either side of the river bed; until these floods subside it is impossible to start work on the erection of a bridge or even the installation of the equipment needed. In normal seasons this meant that only about six or seven months would be available for the erection of the bridge from start to finish; hence it became a frantic rush working twenty-four hours a day to get the work completed before the monsoon broke. Much depended, therefore, on the suitability and reliability of the machinery needed for its erection. Rudyard Kipling's story *The Bridge Builders* gives a vivid, though somewhat dramatised, picture of the difficulties and dangers confronting the engineers in their race for time.

In our London headquarters we had two large drawing-offices, one which dealt with harbours, docks, dams, etc. presided over by Mr. Andros; the other under Mr. Ferreday dealt with bridges and other constructional work needed by railways. Both Andros and Ferreday were masters of their art and were great admirers of my grandfather. Ferreday, in particular, was always kind to me; to him I could always go for advice and encouragement, and to him I spoke of my misgiving as to the adequacy of the equipment under my inspection. He, too, was worried about it.

It became clear to me that to get the sort of specialised equipment we needed we should have to turn to the small general engineering concerns who were prepared to build small numbers, or even individual examples, of any pieces of machinery, provided they were supplied with the designs. Most of these, like Messrs. Lloyd & Plaister, depended for their success upon the personal ability of one individual in supreme command. At that time this country abounded with such small firms, each usually serving a small local area but the difficulty was to find them, for they did not advertise or issue catalogues, nor did their names appear or any 'approved' lists of the India or Colonial Office. Between my visits of inspection to the larger approved firms, I called upon a number of these small firms. Those I visited varied widely, but a few struck me as quite first-rate, ready and agreeable to co-operate, provided they were given at least a scheme design of what was required. One firm in particular, Messrs. Murray Workman & Co., struck me as quite outstanding. They employed between fifty and a hundred skilled mechanics under the leadership of Mr. Murray, then a middle-aged man of wide practical experience, full of initiative and eager to undertake the manufacture of just the sort of equipment we were wanting.

After these researches into the possibility of getting our machinery tailor-made to suit our requirements by some independent firms, I prepared a memorandum urging that I should have a small drawing-office in which to prepare scheme designs for light portable specialised steam, pneumatic and hydraulic machinery. I discussed this first with Ferreday who thoroughly approved, and then put it to my grandfather and his two partners. Though I was now twenty-five years old, my grandfather had never fully realised that I was grown up, but was to some extent re-assured by the fact that the engine I had designed for his car really worked very well. I understood that he agreed, though with some misgivings about my lack of experience. Robertson, I am afraid, had always regarded me as a playboy, too frivolous and light-hearted to be entrusted with such a responsibility; he boggled, too, at my proposal to employ firms not approved by the India or Colonial Office. To this latter objection my grandfather's reply was to the effect 'be damned to official approval – it is our job, as consultants, to supervise the construction of bridges, and we must insist on having a free hand'. Tritton, I believe, agreed with my memorandum, or at least raised no objection, and the upshot of it all was that I was to be allowed to have a small drawing-office with two draughtsmen of my own choosing, but to allay Robertson's misgivings, my designs had to be approved by Ferreday. To this I had no objection for I had the greatest respect for Ferreday's judgement and advice and he, I think, respected my mechanical engineering experience. I was delighted, for it promised the fulfilment of my ambition to start up a mechanical engineering department in the firm under my own wing.

A year or two before this I had tried to persuade my grandfather to let me go to India and take part in the erection of one of his bridges in order to gain first-hand experience with the mechanical equipment then in use. He would not hear of it. At that time cholera, dysentery and malaria were taking heavy toll, and he felt that I had no experience of how to take care of myself in a tropical country, nor was I to be trusted to take the necessary precautions. I had to rely, therefore, on the reports brought back by those members of our firm who had been so engaged. They complained that much of the equipment we were sending out was unsuitable, and some quite useless. In the frantic rush for time, the crying need was for the lightest possible power plant, so designed that it could readily be broken down into separate sub-assemblies for transport by ox-drawn waggon and then re-erected on the site without the need for foundations.

In those days power transmission by electricity was not acceptable because of the failure of insulating materials to withstand hot and humid conditions. All power transmission, therefore, had to be by pneumatic or hydraulic means, chiefly the former, and all power generation by steam engines driving compressors or hydraulic pumps erected as soon as the flooding subsided, and as near to the riverside as possible. In some cases the foundations of the bridge piers had to be laid below a depth of water which might amount to as much as thirty feet, and to make this possible caissons had often to be used. To make conditions tolerable for those doing heavy manual work inside the caisson, a very large flow of air was required at a pressure of a little in excess of the hydrostatic head, the usual pressure range being from 10 to 20 lb. per square inch. Hence, for the first stage in the construction of a bridge the need was for compressors delivering a large volume of air at low pressure, but as soon as the piers were completed, air at a pressure of about 80 lb. per square inch was required to operate the many hundreds of pneumatic tools, such as drills, reamers, chisels and rivetting hammers, for the erection of the steel spans.

Hydraulic power was required for hoists, cranes and jacks and, where possible, for rivetting, and this was transmitted largely through flexible tubing. On that account the pressure had to be limited to about 500 lb. per square inch. In England at that date electricity had virtually superseded hydraulic power transmission over any considerable distance, and where hydraulic power was still in use in this country, much higher pressures of the order of 2000 lb. per square inch were usually employed. All equipment such as hoists, jacks and rivetters was designed to operate at this high pressure. With the low pressure to which we were condemned we had to employ equipment far larger and heavier than was otherwise needed.

It seemed obvious to me that by employing small hydraulic intensifiers interposed between the pipe-line and the jack or rivetter to boost the pressure from 500 to 2000 lb. per square inch, we could get away with very much smaller, lighter and more manœuvrable equipment. Almost my first adventure in my new drawing-office was to design a suitable small and light intensifier consisting of a differential piston and hand-operated valves which could be brought into action as and when needed as, for example, when starting from rest a heavy mass, or for the final closure of a large rivet. There was, of course, nothing new or original in the use of hydraulic intensifiers; for many years past they had formed an integral part of large hydraulic presses. All I could claim was that my first version was such as could be fitted to any individual piece of hydraulic equipment. I got a small but enterprising firm of hydraulic engineers to make up a sample prototype which, after a few modifications, worked very well, and thereafter they made several hundred more for use in bridge erection and other such purposes.

As to pneumatic equipment, the need at that time appeared to be for fresh thinking about the kind of power plant required. In my grandfather's early days the conventional form of steam engine had been the horizontal open type, and the same applied to air compressors, but by the turn of the century their place had been taken by the vertical totally-enclosed high-speed engine and air compressor, sometimes embodied as a single unit, sometimes with engine and compressor direct-connected and mounted on a large cast-iron bedplate. Because of the high speed and imperfect dynamic balance, these machines required massive foundations such as we could not possibly provide on the site in India. Again, their monolithic construction prevented their breakdown into component parts for transport in undeveloped countries. With the increasing use of pneumatic tools, the problem of how to supply sufficient compressed air was becoming acute.

What we needed, I thought, was a combined steam engine and

air compressor which could readily be broken down into component parts, no one of which would weigh more than a ton and which could be assembled on any reasonably level piece of ground without the need for foundations. The air compressor cylinder, together with its piston rod, would form a separate assembly and I planned to provide, as an alternative, two such cylinders, one of very large diameter for the supply of low pressure air for caisson work; the other, of much smaller diameter, to provide high pressure air for pneumatic tools. Either of these otherwise interchangeable cylinders would be mounted in tandem with the steam cylinder and coupled to the protruding end of its piston rod by means of a simple muff coupling. I planned also to provide as a third possibility a double-acting hydraulic pump mounted and coupled in the same manner.

As to the steam engine itself, I proposed to go back to the fashion of half a century earlier and employ a completely open horizontal engine with a single double-acting cylinder.

The main frame of the engine consisted of two parallel lengths of channel iron, each about twenty-five feet in length and spaced about three feet apart. On these were mounted the crankshaft assembly and flywheel as one unit, the steam cylinder with its crosshead and guide as a separate unit, and the air compressor at the rear end of the steam cylinder as a third unit. The main assembly comprising the crankshaft bearings, the crankshaft itself, and the large diameter but light spoked flywheel could readily be broken down to three separate units, each weighing well under one ton. For the sake of lateral stability I provided a third length of channel iron attached at right-angles to the main frame in the form of the letter 'T', to either end of which was attached a foot of the spade and shield type similar to that then in use for anchoring field guns. The third such foot was fitted at the forward end of the main frame; thus the whole machine rested on three widely-separated anchors, and no foundations were needed.

The design of this power plant was, I think, the most important and certainly the largest piece of individual equipment for which I had been responsible but the major share of credit for its success should go to Mr. Murray and the very able foreman of Murray Workman, both of whom entered so whole-heartedly into the project, and contributed so much from their vast fund of experience; in short, they made it work. We carried out a number of very convincing demonstrations in the presence of those members of our staff most concerned with the responsibility of bridge-building, and of officials of the India Office, the Bombay Baroda, East India and other railway companies, the upshot of which was that a first order was placed with Murray Workman by the East India Railway Company for six such power plants for the erection of a very long bridge then being designed by our firm, and four more for the Bombay Baroda Railway Company.

During the year 1911 it was becoming obvious that Robertson was a very sick man. Though I did not know it, he was suffering from cancer of the throat and knew that he was doomed. He had confided this to his partners and urged that they should find someone to succeed him, which they did in the person of a Mr. Frederick (later Sir Frederick) Palmer, then Chief Engineer of the Port of London Authority. Palmer, I was led to believe, was essentially a business man, a go-getter and a first-rate administrator. I cannot remember exactly when he took over but it must have been in 1911 or 1912. Poor Robertson carried on as long as he could but he was often in great pain and died in December 1912. The name of the firm was then changed to Rendel, Palmer & Tritton, and so it remains to this day.

Palmer, it seemed, had known my grandfather for many years and had a great admiration for him, but I wondered what he would think of me, a mere mechanical engineer. Both my grandfather and Robertson took a very poor view of mechanical engineers in general; in their eyes the civil engineer was the artist, the other, a mere mechanic belonging to a much lower position in the hierarchy. On my first meeting with Palmer my doubts were dispelled for he told me that he believed as time went on the civil engineer would become more and more dependent on the mechanical. He encouraged me to continue with my mechanical equipment designs, and said that he looked to me to take charge of that side of the firm's activities. During the remainder of my time with the firm I got on well with Palmer whom I came to like very much, and I think he liked me. Like my grandfather he was no respecter of persons or of government red-tape, and I think that his leadership did much to restore our prestige in the eyes of Whitehall.

Several events occurred during those last years before the outbreak of the First World War which profoundly affected my future career. First, and from my point of view the most important, was that I got married. In the summer of 1909 I had met for the first time, Beatrice Hale, the daughter of our long-standing family doctor and friend. Her father had attended my mother when I was born and had looked after my two sisters and myself ever since. Beatrice had been an art student at the Slade School and was sharing a studio with two young women friends, but she was also helping my aunt, Edith Rendel, and my cousin, Leila, with their welfare work. She was a keen and accomplished dancer, and I must have been hard hit indeed for on her account I endured the agony of a course of lessons in ballroom dancing. During the winter that followed I met her several times at dances and took her for drives in my Dolphin car. In the autumn of 1910 we became engaged, and in June, 1911, we married. After a reception at her parents' house, we set out for our honeymoon in my two-seater car, and on our return took lodgings in a small house beside the river at Walton-on-Thames, from which I used to drive up daily to the London office. The weather during the summer of 1911 was the finest and warmest I can ever remember. On my return from town, we used to go swimming in the river or sometimes to Brooklands Racing Track for a spin at high speed round the course, and generally finishing up with a picnic supper in a punt on the river. There must have been some wet days but my recollection is that throughout the whole of July and August that year the sun shone from a cloudless sky.

In the autumn of 1911 we took a small flat on the south side of the river alongside Hammersmith Bridge. It was not a good site to choose for it proved to be a centre of fogs, too far from the centre of things in London, and too far from the country and navigable parts of the river. In July, 1912, our eldest daughter was born; soon after that event we rented a house and garden at Walton-on-Thames, and there we lived for the next six years.

It had always been my intention to carry on with the research that Hopkinson had initiated at Cambridge, but so long as I lived in London this was impossible. I had, however, collected together a good deal of equipment for use when I could have a workshop of my own in which to carry out engine tests. As soon as we were established at Walton I set up, at the far end of the garden, a large portable wooden shed to serve as a workshop and test shop in which I installed my 5-inch lathe and the now disused pumping engine from Graffham. To enable me to do this, my grandfather had given me a present of £100 and, in addition, royalties from both Lloyd & Plaister and the Britannia Co. were just beginning to come in. These I spent on my new workshop and Hetherington insisted that his share of the royalties be dedicated to my research. In addition to my other equipment I installed a Shoreham-built two-cylinder Dolphin marine engine which I had acquired after the winding up of the Two-Stroke Engine Company. This engine, unlike our later version, had separate detachable combustion chambers which made it an ideal unit for research. Most valuable of all, I still had Hopkinson's precious optical indicator; thus I was admirably equipped to undertake a research into the phenomenon of detonation which Hopkinson had recommended me to follow.

At that date, 1912, it was still the almost universal belief that knock in the petrol engine was due to pre-ignition, initiated by some hot surface within the combustion chamber. Hopkinson did not accept this; he had suggested to me that the knock was due to the setting up of an explosion wave within the combustion chamber of sufficient violence both to set the walls of the chamber in vibration and to dislodge or shatter the mirror of our optical indicator; thus, while at Cambridge, I never succeeded in getting a satisfactory indicator diagram from a knocking engine.

In the years that followed I had succeeded in attaching the mirror of my indicator more securely so that it could withstand the shock of impact of an explosion wave, if it were not too severe. The Dolphin engine I had installed in my workshop was prone to knock even though its compression ratio was well below 4 to 1. On this engine I was able, for the first time, to see and to photograph indicator diagrams under knocking conditions; from these it was apparent that the rate of pressure rise following ignition by the sparking plug was perfectly normal until almost the end of the combustion period, but at the last moment it shot up vertically, thus setting up some vibration which persisted during the ensuing expansion stroke, giving the expansion line a jagged sawtooth appearance. From these diagrams, and from much other corroborative evidence, I came to the conclusion that the knock in the petrol engine was caused by the spontaneous combustion or detonation of some small part of the working fluid due to compression by, and radiation from, the rapidly advancing flame front. Whether or not this would happen depended, to a large extent, upon the opportunity the as yet unburnt charge had of getting rid of its heat to the surrounding walls before its temperature was raised to that of spontaneous ignition. What the temperature might be would, I thought, depend upon the chemical and physical characteristics of the fuel about which I was abysmally ignorant.

At about this time benzole derived from coal appeared on the market as an alternative fuel, and with benzole I could get no trace of detonation under any circumstances. I made a new piston and combustion chamber, raising the compression ratio to just over 5 to 1 and thereby scoring a gain in power output of about twenty per cent, still without any trace of knock. This, to me, was rather a startling observation.

I had recently made friends with a certain Dr. Ormandy, a consulting organic chemist, and expounded to him my theory as to the mechanism of detonation or 'knock' and of my rather startling observation of the behaviour of benzole. He said he was not surprised, for commercial benzole was a hydrocarbon of the aromatic series whose molecule was of ring formation and, as such, very stable; going on to say that commercial petrol was composed of a mixture of hydrocarbons of the paraffin, napthene and aromatic series, that the paraffin series predominated and that the molecules of this latter series were all of loose chain formation and, as such, much less stable. He very kindly offered to prepare for me some small samples of pure substances of paraffins, napthenes and aromatics. These I tried out in my Dolphin engine and found, as Ormandy had predicted, that so far as their tendency to detonate was concerned the paraffins were by far the worst and the aromatics the best of all. Thus, by the year 1913 I had satisfied myself that the incidence of detonation was the most important factor limiting the compression ratio and therefore the power output and efficiency of the spark-ignition engine, and that it was a phenomenon entirely distinct from pre-ignition.

When a little later I had completed the single-cylinder version of my supercharged aero-engine I found that it could serve as a valuable instrument for the knock rating of fuels and this stood me in good stead a few years later.

Another very important event for me occurred during that period. One day during 1913 a tall man with a broad Scots accent called on me at my office and introduced himself as Mr. John Blackie, senior partner in the publishing firm of Blackie & Sons of Glasgow. He told me that he and his partner considered that the time was ripe for the publication of a new book on the internal combustion engine, and that the purpose of his visit was to invite me, on Hopkinson's recommendation, to be the author of such a book. He went on to say that Dugald Clerk's book *The Gas Engine,* first published in 1886, which subsequently appeared as *The Gas, Petrol and Oil Engine* (1909) was still regarded as the classic work on the subject, and although its author had revised and added to it, its run as a best seller could not last much longer. Of the many other more recent books on the subject none, from the publisher's point of view, had proved a success. He went on to say that in his opinion what was wanted was a new book on much the same lines as Dugald Clerk's, that is to say, attaining a nice blend of theory and practice, couched in language that all could understand, and in readable English.

This invitation rather took my breath away and I pointed out to him that I was a mere amateur; that I had never for one moment dreamt of writing a book, nor had any idea as to how to set about it, and that there must be many better equipped than I to undertake the task. He was not deterred but left saying he hoped that, after thinking things over, I would comply with his request, and that if so he would give me a letter stating that his firm had commissioned me to write a new book on the internal combustion engine which I could show to any firm I visited, and which he hoped would prove an 'open sesame' to any sources of information I needed.

Sir Alexander Rendel   Halsey Ricardo

Harry Ricardo   Beatrice Ricardo, née Hale

Pumping Engine, c. 1902

Model Steam Engine, c. 1897

'The Model Steam Engine that did not look in the least like a Steam Engine.'

CENTIMETRES

0 1 2 3 4 5 6 7 8 9 10 11 12 13 14 15

I sought out Sir Dugald Clerk who told me that he considered Blackie & Sons to be a first-rate firm. He said that in his experience there was nothing like writing such a book for bringing out and forcing one to fill in the gaps in one's knowledge, and his advice was therefore to go ahead and accept the offer, and so I did.

At my next meeting with Mr. Blackie we agreed that the work should be in two volumes, one dealing with heavy industrial engines, such ts gas, oil and diesel engines, and the other with light, high-speed petrol engines, the latter involving an entirely different conception of mechanical design. Up to this point neither of us had mentioned the matter of terms, but just before parting Mr. Blackie announced that his firm would prefer to pay in a lump sum rather than a continuing royalty, and that they were prepared to pay me the sum of £500, half of which was to be paid when the manuscript was received, and the remainder when the book was published. This was a far greater sum than I had ever expected, for it amounted to considerably more than a year's salary.

I had made a start on the book and was getting on nicely with the first volume when the 1914 war burst upon us, and all our plans were changed.

CHAPTER 9

The End of the Beginning

Earlier in the present book I have given to the best of my recollection my impressions of the state of the art at the turn of this century. Now I will try to recall the state of the art at the outbreak of the First World War, and the changes that took place during the intervening thirteen years. Those years saw the almost complete extinction of the piston steam engine as a means of power production on land, its place being taken by the steam turbine for large powers, such as electric power-stations, and by the internal combustion engine for small and medium powers, though it still retained a dwindling foothold in under-developed countries, such as India. While the power output of the piston steam engine had been limited by the sheer magnitude and weight of its moving parts to something of the order of 10,000 h.p. from a single crankshaft, no such limitation applied to the turbine. By 1914, if I remember right, turbo-electric generating sets, each of 100,000 h.p., were being talked of.

At sea the steam turbine was rapidly replacing the piston engine in all the larger and faster ships. The famous Cunard liners *Lusitania* and *Mauretania*, each of 35,000 tons, and with a service speed of 27 knots, marked an epoch in that period.

During the first decade of the present century the Navy had adopted the steam turbine for the propulsion of all its larger and faster ships, early examples of which were the battleship *Dreadnought* and the destroyers *Viper* and *Cobra*. In this connection I remember well while in my early Cambridge days being taken around his shipyard at Chiswick by Sir John Thornycroft who showed me, among other things, a pair of engines, one complete, the other partially erected. They were four-cylinder, triple-expansion, narrow-angle vee engines of extremely light construction. To this day they remain in my mind as one of the most beautiful examples of mechanical design I have ever seen, and of which

130

he was justly very proud. He told me that they were destined for a new torpedo-boat then being built at his new shipyard at Southampton; that each would develop 3000 h.p. at a speed of, I think, 300 r.p.m. With a sad note in his voice he said that they represented the last piston steam engines that he would ever build for the Navy; that his latest destroyer, the *Viper,* then under construction at Southampton, would be fitted with Parsons' turbines estimated to develop 20,000 h.p. and with which he hoped to reach a speed of 40 knots.

The huge increase in shaft horse-power which the turbine offered could be realised only in conjunction with the newly-invented Michel thrust block and with the burning of oil fuel in place of coal.

The steam locomotive remained almost the last stronghold of the piston steam engine which in all its fundamentals had undergone little change during the period under review, but the increasing weight and length of both passenger and freight trains demanded a corresponding increase in boiler capacity. To this was added the demand for the steam heating of passenger trains, then an innovation. This caused the locomotive boiler to swell in all directions which the railway loading gauge would permit, in height until the funnel became a mere sawn-off stump and the steam dome a flattened mound, in depth until it became necessary both to reduce the size of the driving wheels and to banish the valve mechanism from under the boiler to outside the frames. No longer did the locomotive remain a thing of beauty and elegance, the pride of its driver and of an admiring audience at every station; instead it had lost its figure and turned into a bloated overfed monster lacking in dignity and self-respect. The elegant ballet girl of the nineties was growing into the corpulent and untidy frump of the present century. My father, with his love of beauty and keen sense of proportion, abhorred the locomotive's aesthetic downfall, and so did I.

I do not know whether by this date the rapid development of mechanised road transport had begun seriously to affect the railways, but the writing must have been on the wall. As far back as the nineties, my grandfather had pointed out to Whitehall that mechanised road transport would always have the advantage in that it could deliver passengers and freight from doorstep to doorstep, which, of course, the railways could never hope to do. On the other hand, the railways would for long reign supreme in the transport of heavy freight over long distances; hence future planners should concentrate on integrating the two forms of transport, that on the road as feeders to the railways rather than in competition with them.

The advent of the steam turbine made possible the establishment of large central power-stations in our chief cities, each serving a wide area, throughout which alternating current at a uniform voltage and a frequency of fifty cycles per second could be relied upon. With this assurance of uniformity, the small electric motor for domestic and other uses at once became a practical proposition. It soon displaced the hot air engine and the smaller sizes of the gas and oil engine.

Plans were already in hand both for linking together these large central power-stations and spreading an electrical grid over the entire country, but this ambitious project was not completed until the early 1930's. In the meantime, the smaller provincial towns each had electric generating stations of their own powered, for the most part, by diesel engines, as did the large factories out of reach of the existing public supply. As the coming of the steam turbine ended the reign of the piston engine, so the coming of electricity, derived from the steam turbine, brought the reign of the gas engine to a close, but not that of the diesel engine.

The rapid development of the automobile during this period had led to a large demand for petrol; there was also a large and growing demand for kerosene as the poor countryman's source of light and heat, but there was little demand for the heavier distillates, such as gas oil, which were becoming a drug on the market, as also was creosote, a by-product from the distillation of coal. The diesel engine, unlike other internal combustion engines, could digest these fuels, and that, too, at a higher efficiency than any other heat engine. It afforded, therefore, the cheapest source of power for industrial use. Owing partly to the very high cylinder pressures necessary to ensure ignition, and partly to the need for still higher air pressures both to pulverise and distribute the liquid fuel throughout its combustion chamber, it was costly, complex and very heavy.

During the period under review every effort was being made both to reduce the weight of the diesel engine and to dispense with the need for blast air, which involved the addition of a multi-stage air compressor, in itself a complex, bulky and expensive component. Theoretically the diesel engine operated on a constant pressure cycle and, so long as blast air was employed, this could be achieved in practice, thus limiting the maximum cylinder pressure to something under 500 per square inch. As I saw it, the blast air at a pressure of about 900 per square inch played three important functions: firstly, so thoroughly to pulverise the liquid fuel as to reduce it to a fine mist, secondly to distribute that mist uniformly throughout the combustion space, and thirdly to set up very violent turbulence and thus accelerate the spread of combustion. With the help of blast air the delay period before ignition took place was reduced to the absolute minimum. Hence, by controlling the rate of admission of the liquid fuel, the cylinder pressure could be held almost constant until the end of the injection period, but the addition of a high pressure air compressor absorbed between 5% and 10% of the useful power, which was a high price to pay both in first cost and overall efficiency. At some time during this period, McKecknie of Vickers had designed and developed a very compact high-speed diesel engine for the propulsion of submarines for the Navy and rumour had it that he had succeeded in dispensing with blast air injection, but the whole development was so closely veiled in secrecy that little or nothing was known about the performance of these engines until after the outbreak of the First World War, when other firms such as Mirrlees and Crossley Brothers were called upon to build such engines to a Vickers' design. It then transpired that Vickers had got away from blast air injection by spraying the fuel at an enormously high pressure through a pepper castor nozzle perforated by a large number of very minute orifices, through which the liquid fuel was projected at high velocity into all parts of the combustion chamber. Others before him had, of course, attempted to do the same thing, but McKechnie's success was, I think, due to his boldness in employing far higher fuel line pressures and to his ingenious design of oil pump, and to the use of hydraulic accumulators in the form of Bourdon tubes attached to each injector. Above all, success was due to the excellent mechanical design of his engine as a whole.

When at length performance figures became available, it appeared that the maximum power output with a reasonably clean exhaust was inferior to that of an air injection engine, and this despite the fact that he had been forced to depart from the constant pressure cycle and accept cylinder peak pressures of well over 600 per square inch. It was evident that the lower brake mean pressure of the Vickers engine was due to poor utilisation of the relatively stagnant air in the combustion space.

Apart from this handicap, which affected only its full power performance, the Vickers submarine engine gave excellent service throughout the whole of the 1914–18 war, and our naval engineers were thankful to be rid of the blast air compressor, and of all the additional plumbing it involved.

While Vickers were engaged in developing an airless injection diesel engine, Hesselmann, then Chief Engineer of the Atlas Company in Stockholm, who were building submarine engines for the Swedish Navy, had also been tackling the same problem, but from a different approach. Hesselmann argued that so long as the air in the combustion space was relatively stagnant, it would be impossible, without the aid of blast air, to bring the fuel into intimate contact with it, and that, in his view, the right course was to set the air to find the fuel rather than the other way about. By masking part of the circumference of the air inlet valve, he succeeded in setting up a rapid rotary swirl of the air round the axis of the cylinder, a swirl which persisted throughout the compression stroke. His injection nozzle had only four relatively large orifices set at 90° to one another and from each of these four orifices the fuel was projected radially outward as a coarse spray. The rotating mass of air would sweep past these fuel jets, and thus a greater proportion of it would be brought into contact with the fuel than would be possible if the fuel was set to find the air. By adopting this method he was able to get away with a much simpler and relatively low pressure fuel injection system.

Hesselmann made no secret of what he was doing; he invited

me to visit him while I was in Stockholm in July 1914, when he showed me his latest submarine engine on test. From his test figures it was obvious that he was getting a brake mean pressure greater than that with his air injection engine, and a somewhat lower specific fuel consumption, but to achieve this he had to accept considerably higher peak pressures. At that time mechanical engineers generally, and in this country in particular, had a horror of high cylinder pressures, borne of past experience with steam piston engines, and were apt to purse their lips at the suggestion of anything higher than 500 per square inch. Hesselmann, however, did not share this squeamishness; he appeared quite content to let his peak pressures rise between 600 and 700 per square inch, and thereby reap the gain due to the more efficient heat cycle.

I had long cherished the hope that some day we might be able to produce a light, compact and high speed diesel engine for road transport, but so long as we were wedded to blast air injection this would be out of court on grounds of cost, bulk and complexity. After seeing what Hesselmann had achieved, my dream of the distant future had been transformed into an immediate practical possibility but the First World War broke out only a few days after my visit to Hesselmann, and banished all such thoughts from my mind for several years to come. When, however, some ten years later we set out in earnest to design and develop a really light high-speed diesel engine, it was Hesselmann's teaching that we followed.

While the period under review in this chapter saw the extinction of the piston steam engine and of the gas engine and the waning of the vaporising oil engine, it saw also the rise of the hot bulb or so-called semi-diesel engine. This was really a logical development of the Hornsby-Ackroyd system, the important difference being that instead of using the hot bulb as a vaporiser for fuel injected into it at an early stage during the suction stroke, a much higher ratio of compression was employed, and fuel was injected under pressure during the latter half of the compression stroke. In the former case, most of, if not all, the fuel was in the vapour phase at the time of its ignition; in the latter most of it was still in the liquid phase and combustion took place from the surface of the droplets as in a full-blooded cold-starting diesel engine. By virtue of its fairly high compression ratio, usually of the order of 7 or 8 to 1, as compared with about $3\frac{1}{2}$ to 1 in the Hornsby-Ackroyd, the specific fuel consumption of the semi-diesel was relatively low.

This form of engine became very popular. Compared with the full-blooded diesel engine, it had the advantages that it was much cheaper to build, it required no blast air and its fuel injection system was of the simplest, and it could digest the same cheap fuel. Its disadvantages were that it took a long time to heat up the bulb before starting and the uncooled bulb, when subjected to high pressure and thermal stresses, was liable to crack. Also, on light loads or when idling it did not keep hot enough and hence, like the vaporising oil engine, there was only a limited range of load over which it would operate efficiently. Despite these disadvantages it was very popular, more especially as a marine engine for fishing boats and so forth, for which purpose it was generally made in the simple two-cycle form with crank-case compression. Although once the cold-starting diesel had learned to dispense with air blast injection and to tolerate higher peak pressures the semi-diesel was doomed, in 1914 it was, I think, at the zenith of its popularity.

To me by far the most exciting development during this period was that of the petrol engine, both for road transport and for aircraft propulsion. In its broad essentials the petrol engine differed from the gas engine only in the use of a different gas, but it entailed an entirely different conception of design and manufacture, untrammelled by tradition or prejudice. In the stationary engine field, designers were content to operate at quite low revolution and piston speeds, hence dynamic problems such as inertia of the moving parts did not trouble them, nor were they much concerned with the weight or bulk of their engines. Their main objective was to achieve both a high thermal efficiency and a long working life and to them a speed of over 300 r.p.m. was regarded as a dangerous adventure, even for quite a small engine. In the automobile field, reduction of weight and bulk became of first importance, and to achieve this it became necessary to operate engines at revolution speeds undreamed of hitherto. Even before the close of the last century, small motor-cycle engines were running at speeds of 2000 r.p.m. and over. This breakaway from all past tradition called for an entirely fresh conception of mechanical design; light moving parts and dynamic balance became of first importance, stiffness rather than strength ruled the day, inertia and centrifugal forces became of more importance than peak cylinder pressures, and so on.

In my memory, the first five or six years of the present century was a period of groping and selection during which many different types and designs of petrol engines for road service were tried out. During the next few years most of the more exotic types had fallen by the wayside; the four or six-cylinder engine had become the most popular form for the larger cars and commercial vehicles, and the single-cylinder for light voiturettes or cycle-cars. Attention became focussed on improving the manners and the performance of the prevailing types, the former with a view to reducing noise and vibration and the latter to increasing the power output and fuel economy.

During the earlier years of this period, it became the fashion to employ a short stroke, usually about equal to the cylinder bore, in order to enable such engines to operate over a wide speed range without too greatly exceeding the then conventional limit of piston speed of 1000 feet per minute. In my design of the Dolphin engine as made at Shoreham, I followed the prevailing fashion and made the bore and stroke equal.

In a lecture I had attended in, I think, 1910, Pomeroy of Vaux-halls argued the case for a long stroke engine. He contended that piston speed was limited only by breathing capacity, and that in his view there was no reason whatever why piston speeds up to 2500 feet per minute should not be obtainable with a single row of valves, and of 3000 feet per minute or even over if a double row be employed. He went on to argue that the piston itself was the most vulnerable member in an engine on account of the high thermal and pressure stresses to which it was subjected, also that its weight must be kept to the minimum in order both to reduce inertia forces and, in the case of four-cylinder engines, vibration due to the unbalanced secondaries. Hence, he argued, it was desirable to keep the piston as small as possible. He pointed out that by doubling the stroke of the piston it would generally be possible to double the power output of contemporary engines and still keep within the limiting piston speed dictated by breathing capacity without increasing the rotational speed which was then controlled by the dynamics of the valve mechanism, crankshaft torsional vibration and rising friction. At this lecture he threw on the screen a photograph of a beautifully designed feather-weight steel piston which he was using in the Vauxhall racing-cars. I was tremendously impressed by this lecture which struck me as a model of lucidity, and the first attempt I had heard to rationalise the design of the petrol engine. Shortly after this the famous 30–98 Vauxhall car made its appearance on the racing-track at Brooklands. Fitted with a small bore long stroke engine of 4.625 litres capacity, and developing just over 100 h.p., the car achieved a speed of over 100 miles per hour and, driven by Kidner, it could keep its end up in competition with the Napier, Fiat and Mercedes cars with their giant engines of more than three times its total cylinder capacity. Henceforth, and over the next twenty-five years, it became the fashion to employ small bore long stroke engines and Vauxhall's example was quickly followed by Sunbeam with a three-litre engine of 80 mm. bore and 150 mm. stroke. On the Continent it became the fashion to employ even more extreme stroke-bore ratios; for example, the 1911 Hispano-Suiza engine had a cylinder bore of 80 mm. and a stroke of 180 mm., while Peugeot went even further and adopted a stroke of 200 mm. for the same cylinder bore.

In these early years little attention had been paid to the detail design of cams for operating valves, with the result that valve mechanisms were very noisy. This had been tolerated until the Knight double sleeve valve engine appeared on the scene and was adopted by the British Daimler Company and several continental firms. The appearance of this almost perfectly silent engine created quite a sensation and set the designers of poppet-valve engines looking carefully into the dynamics of their valve mechanism and, for the most part, they found them wanting. I remember about that time attending a lecture by Pomeroy on engine noise generally and on the dynamics of valve operation; in this lecture he advocated the employment of a gradual approach ramp in order to take up the clearance between the valve and its tappet. By this and certain other features which he described he certainly succeeded in evolving a very silent valve mechanism. Whenever possible I attended Pomeroy's lectures and sometimes took part in the discussion on them, as a result of which he invited me to the Vauxhall works where, at lunch, I met also Percy Kidner, then Managing Director who, twenty years later, became a member of the board of my company. This visit to Luton was the first time I came into personal contact with any of the leading members of the automobile industry.

During the earlier part of this period the steam car was a formidable rival. Compared with the petrol-engined vehicles of the time, it had the merit of being almost completely silent and vibrationless; it required no change speed gears while the thermal storage capacity of its boiler gave it a reserve of power for acceleration, hill-climbing or bursts of speed which the petrol engine could not achieve. On the other side of the balance sheet the steam pleasure car was open to the objection that despite its makers' claim to the contrary, it took a long time to get up steam, its consumption of water was heavy, namely, about 2 miles per gallon, and the fire risk was much greater. From the Continent came such examples as the Serpollet and the Miesse, both large and heavy cars equipped with flash boilers generating high pressure highly superheated steam and with single-acting poppet-valve engines. From America came the Stanley and the Locomobile, both small and very light cars equipped with very short vertical fire tube boilers supplying medium pressure wet steam to very neat, cleverly-designed two-cylinder double-acting engines. These were followed a little later by the American White steam car with so-called semi-flash boiler, and a double-acting compound steam engine. In general, the continental steamers used kerosene as fuel, and the American petrol; with the latter control of the burner was easier, but the fire risk greater. In some cases condensers were fitted to save water, but these soon became inoperative due to oil carried over by the exhaust steam and to clean them out was an augean task. Because of its disabilities the steam pleasure car never became a utility vehicle but it lingered for many years in the hands of enthusiasts.

In the commercial vehicle field, the battle of steam versus petrol was still in full swing at the outbreak of the First World War. By then petrol had won the day in the lighter classes of vehicle, but steam still retained its hold in the heavy lorry class. Such firms as Thornycroft, Foden, Clarkson and Sentinel were all busily engaged on the manufacture of large coal or coke-fired steam lorries for heavy goods haulage and with these the petrol engine could not yet compete. So far as I can remember, the first self-propelled buses in London were all steamers, some made by Thornycroft with water-tube boilers, and others by Clarkson, with his very efficient thimble-tube boiler. All were double-deckers and had tall funnels extending well above the heads of the passengers on the upper deck and were a strange sight in those Edwardian days. They had wooden wheels with solid rubber tyres for the large pneumatic tyre had yet to come.

In the autumn of 1903 the Wright brothers in America startled the world by their achievement of sustained flight with a heavier-than-air machine. For this they employed a small but conventional four-cylinder, four-cycle petrol engine of their own design. This spectacular achievement caused great excitement in France, Germany and Italy but, according to my recollection, very little in either this country or America, so little in fact that to acquire the financial aid they needed for further development, the Wright brothers had to turn to France, where they gave a number of demonstration flights. Immediately the leading designers and manufacturers of petrol engines in that country turned their attention to the development of very light petrol engines for aircraft propulsion. Many and various were the designs they produced during the next few years, all bristling with ingenious and original ideas, the most outstanding and daring example being the famous Gnome engine in which the radial air-cooled cylinders rotated around a stationary single-throw crankshaft. Looking back over the years I am amazed at both the daring and originality of the designer, Monsieur Seguin, who cast aside all conventional tenets of mechanical design. At the time of its birth the Gnome was by far the lightest engine for its power ever produced and so it remained for many years to come. The success of the engine stimulated other French designers to develop rotating cylinder engines, such as the Le Rhone and Clerget which, though basically similar, differed in many important details of design.

Of the many other French aero-engines developed during those early years I recall in particular the Antoinette, an eight-cylinder water-cooled vee engine, a masterpiece of more conventional design. In it, too, weight saving had been carried to the extreme limit and it bristled with ingenious features; for example, in place of the conventional carburettor its fuel was metered and sprayed into each individual inlet port by a pump. This little engine, weighing less than 200 lb., was said to develop 60 h.p. It was used by several of the early pioneers of flight, among them my school friend Moore Brabazon (later Lord Brabazon of Tara) who told me that it was the smoothest running and most easily controlled of any engine he had flown with during those early Edwardian years. Another French-designed aero-engine was the Anzani air-cooled radial engine with fixed cylinders and rotating crankshaft. This was produced with varying numbers of cylinders from three to seven, each cylinder developing about 8 h.p. It was with a three-cylinder Anzani engine that Bleriot flew the English Channel in 1909.

Of other French aero-engines developed during those early years I recall the air-cooled twelve-cylinder Renault engine which developed well over 100 h.p., and this was, I think, the best and most powerful of the French engines developed prior to the First World War. There was nothing particularly novel about it except that, unlike the other early French engines, its propellor was driven through a reduction gear, thus allowing a higher crankshaft and piston speed.

During a visit to Germany in 1911 I took the opportunity to find out as much as I could about the design and development of aero-engines in that country. There I found a very different approach to the problem. While the French were going all out to produce the lightest possible engine, the Germans were concentrating rather on fuel economy and mechanical endurance. While almost all the French engines were of something less than 100 h.p., the German engines, on the other hand, appeared to be all in the range of 150 h.p. to 250 h.p. While the French engines displayed a wide variety of types and cylinder arrangements, all the German engines I saw were of one general form, namely, six-cylinder, vertical-in-line engines. I saw such engines under construction at the firm of Benz at Mannheim, of Mercedes at Stuttgart, of Körting at Hanover and of B.M.W. at Munich. All had aluminium crankcases split on the centre line of the crankshaft, and with their crankshaft bearings carried half in the lower half and half in the upper half of the crankcase. All had fabricated steel cylinders and all, except the Benz engine, had overhead camshafts and hemispherical combustion chambers. Of those I saw the Benz engine alone had vertical overhead valves, pushrod operated from the camshaft in the crankcase. I was surprised to see so many engines in course of construction by the leading manufacturers in Germany but I was not altogether surprised, therefore, to be told in 1914 when war broke out that Germany had at least 500 fully-developed military aircraft at her disposal.

During this visit I found the Germans most forthcoming and perfectly ready to show me what they were doing and how they were doing it. At the Mercedes works, for example, I watched the fabrication of steel cylinders and steel pistons. This was a work of art carried out by highly skilled craftsmen.

I enquired what they were doing about supercharging for high altitude work and they replied that although they realised that this would have to come sooner or later, the need had not yet arisen. They considered that, for the time being at any rate, the best answer to the altitude problem was to concentrate on achieving the highest possible volumetric and mechanical efficiency by cutting down the reciprocating weight to the barest possible limit, and being content with relatively low revolution speed; hence they preferred to mount the propellor direct on the crankshaft.

In this country and in America flying seemed to be regarded as a 'stunt' by all except a few enthusiasts, such as Grahame White, Handley Page, Moore Brabazon, De Havilland, A. V. Roe and a few others who had built, or were building, machines of their own but who had to rely on French engines for propulsion. To the best of my recollection none of the British firms or leading designers of the Edwardian period appeared to have given any serious thought to the development of a really light aero-engine. It was not until 1909, six years after the Wright brothers' achievement, that the public imagination was suddenly awakened by

Bleriot's feat of flying across the English Channel. It was, I remember, with a thrill of horror that we realised that our island fortress might not be so impregnable as we had fondly believed. Hitherto the British public had regarded aviation as a dangerous sport at which a few well-to-do dare-devil young men were prepared to spend their money and risk their lives, but Bleriot's flight changed the whole picture. The daily press organised public demonstrations of flying and competitions for which monetary prizes were arranged; even so, however, the military authorities took no interest whatever in flight by heavier-than-air machines nor was there any incentive to encourage industry to develop a British designed aero-engine.

Both the Army and Navy were interested only in balloons, captive and dirigible, for observation purposes. In 1909, O'Gorman, whom later I came to know very well, became Director of the Royal Aircraft Factory at Farnborough, later known as the Royal Aircraft Establishment. His terms of reference were to carry out research and development work on captive balloons, non-rigid airships and man-carrying kites on behalf of both Services, but no funds were available to him for the development of heavier-than-air machines. O'Gorman, a first-rate administrator, gathered round him a very capable staff of young civilian engineers and scientists including Geoffrey De Havilland, my Rugby school friend Busk, an authority on aerodynamics, and Colonel Cody an authority on man-carrying kites. Thus equipped he set to work on his alloted task, which he carried out admirably, but both he and the members of his staff were great believers in the future of the aeroplane as a military weapon. It was not until several years later that he succeeded in extracting from the War Office a very small grant to be spent on the development of the aeroplane.

Although the larger and well-established firms were not seriously interested in the development of an engine for aircraft, several small firms did try their hands. For example, Mort of the New Engine Co. had built a neat four-cylinder vee type two-cycle engine in which the four working cylinders were scavenged and charged by means of two Roots blowers, one delivering air as a scavenging charge, and the other carburetted mixture. This little engine, of about 50 h.p. was cleverly designed in that use was made of the Roots blowers as dynamic balancers. So far as I can remember, however, only a few prototypes were ever built. One of these was flown by Moore Brabazon and afterwards found its way into the Science Museum in London.

Another interesting engine was a six-cylinder single sleeve-valve four-cycle engine made by the Argyll Motor Co. It was entered for a War Office competition in which it put up a remarkably fine performance, both in the way of high power and low fuel consumption but unfortunately it broke its crankshaft during the trials and was disqualified. At that time the Argyll Motor Co. was in very low water and the development of this would-be aero-engine was their last throw before going into liquidation.

Another British-designed aero-engine of the same period was the Green, a six-cylinder in-line engine of very amateurish design. Its performance was poor due probably to inadequate breathing capacity but it succeeded in gaining a reputation for reliability on the strength of having gained first prize in a competition held by the War Office shortly before the outbreak of the First World War. It was, I remember, flown by Cody, but he discarded it in favour of an imported Austro-Daimler engine of far better performance.

Although the major automobile firms had apparently given little or no thought to the matter, a few such as Wolseley, Sunbeam and E.N.V. had converted their standard motor car engine for the propulsion of aircraft merely by doubling up two of their four-cylinder engines to form an eight-cylinder vee and, at the same time, scraping off a little superfluous metal. I did not think that any of these improvisations could be regarded as serious attempts to develop an engine for aircraft production. In point of fact, with only one single exception, none of these early British-designed aero-engines survived to take any active part in the First World War, the one exception being the R.A.F. (for Royal Aircraft Factory) engine designed and developed by O'Gorman's very able staff at Farnborough, at first in face of, and later with the reluctant approval, of the War Office. This highly successful engine powered most of our aircraft during the earlier phases of the First World War. It was produced both as an eight and as a twelve-cylinder engine developing 100 and 140 h.p. respectively. Its designer, F. M. Green, had left the Daimler Company to join O'Gorman at Farnborough; there he had the assistance of De Havilland, Busk and others on the aerodynamic side. Thus, for the first time in England, both air-frame and engine were developed in conjunction. So successful did the R.A.F. engines prove that during the First World War over 4000 of them were produced by Daimler and other motor firms.

Looking back over those years it seems shameful that we in England should have allowed ourselves to lag so far behind in the development of aeroplanes and engines, but it was lack of incentive, not ability, that brought this about. Both in France and in Germany the military authorities were quick to see the part that aviation would play in future warfare and lavished huge sums on development but the authorities in this country were blind to the possibilities, while industry as a whole was not prepared to gamble on the chances of its commercial use. In after years O'Gorman told me that he got severely ticked-off by the War Office for having 'squandered' over £2000 of his grant for airship development on aeroplanes and engines which would never be of any military value.

In my Cambridge days Hopkinson had told me of his belief in the great future of the aeroplane, and of the crying need for the development of a suitable engine for its propulsion, and it was really to this end that our research was directed. Hopkinson maintained that the aeroplane of the future, whether it be for commercial or military service, must be capable of carrying a really worthwhile pay-load over long distances. Low fuel consumption was of even more importance than low engine weight. He maintained also that a far greater engine power would be needed than so far had been envisaged. He was full of admiration for the skill and ingenuity the French were displaying in the development of small feather-weight engines of around 50 h.p., but he thought that our aim should be to develop a highly efficient engine of 200 h.p. or more, and he advised me to think about the design of an engine of at least this power. During the years that followed I amused myself by trying out many and various scheme designs of an aero-engine of about 300 h.p. Of the many designs I evolved that which I liked best was of a very narrow-angle, twelve-cylinder vee type four-cycle engine. In this I planned to use the underside of my working pistons to provide a top-up supercharge of pure air admitted through ports round the lower end of the cylinder liner uncovered by the piston at the bottom of its stroke. By this means I hoped to achieve a considerable measure of stratification, and by varying the mixture strength admitted to the cylinder in the normal manner, I could at will utilise the supercharge air either as a diluent to reduce fuel consumption at cruising speeds, or as a power boost for high altitude flying. I estimated that in the former case I would be able to reduce fuel consumption by nearly 10%; in the latter I would be able to maintain ground-level conditions up to an altitude of 10,000 feet. I showed this design to Hopkinson and Dugald Clerk; they were both encouraging about it and made an appointment for me to meet a high-ranking officer at the War Office to whom I showed my design and explained what I was aiming to achieve. He listened to me very courteously and then said he was afraid I was wasting my time. He told me that the only use the Army had for aviation of any kind was for spotting for artillery, and that for this purpose an altitude of only a few hundred feet was all that was needed. The captive balloon was a far better proposition in that it could stay still, was silent, and in direct telephonic communication with the guns below, whereas the aeroplane had to keep milling about, was very noisy and had to depend upon wireless which, at that date, was a very uncertain quantity. Apart from these disabilities the aeroplane was, in his view, far too fragile for the rough and tumble of military service. This was somewhat disappointing: none the less I decided to go ahead.

CHAPTER 10

The First World War

In the latter part of July 1914, my wife and I went for a holiday in Stockholm, leaving our two-year-old daughter with my parents. At that time no thought of impending war entered our heads. I had, however, heard a few expressions of alarm at Germany's rapidly growing naval power and of vague troubles in the Balkans which had then just come to a head with the murders at Sarajevo, but except for a few inveterate prophets of gloom the general public seemed to have no inkling of impending disaster. Before starting I had written to Dr. Hesselmann at Stockholm, to Krupp's Naval Yard at Kiel and to Professor Junkers at Aachen, saying that I would much like to visit them, and in particular to hear about the high-speed diesel engines they were developing for submarines. In each case I enclosed a copy of a letter of introduction which Blackie had provided. At Stockholm we had a wonderful holiday sailing around the numerous islands in the Baltic. Hesselmann was most welcoming and showed me all he was doing. While at Stockholm I received a letter from Professor Junkers saying that he would look forward to seeing me at Aachen and to showing me round his laboratory there, but up to the time we left Stockholm I had had no reply from Krupps.

While at Stockholm I had not seen an English newspaper, and letters from home contained no disturbing news of any kind. From Stockholm we went to Berlin for one night, and then on to Cologne where we stayed for two nights. Leaving my wife sight-seeing at Cologne, I went to spend a day with Junkers at Aachen. Junkers at that time was busily engaged on the development of his opposed piston two-cycle engine as a marine diesel engine. I was deeply interested in all he showed me in his magnificent laboratory. While showing me round he was called off to the telephone and on his return he asked me when I was returning to England; I told him the next day and he then said he thought I would be well advised to go back as quickly as possible; that he had just been informed that Russia was mobilising on Germany's eastern frontier, and he shuddered to think what that might portend. That was the first hint I got that serious trouble was brewing. We returned to England next day by way of Ostend. The boat was crowded with English travellers returning hurriedly from holidays in Belgium and they told us that they had been warned to go home at once and that war might break out at any moment. That was I think August 2nd 1914. Next day we heard rumours that Germany was threatening to invade Belgium. On the following day these rumours were confirmed, and we were told that meant that we were at war with Germany. I suppose we must have been very unobservant but at no time during our three days in Germany did we see any sign of military activity or of popular excitement. On our return home I found a letter from Krupps, which had just missed me at Stockholm and been forwarded to my home address. In it they apologised for their delay in replying to my letter because they first had had to get approval from the naval authorities. This approval had now been given, my visit would be most welcome and they would gladly show me all they were doing. The date of this letter was, I think, July 27th; I mention it because it has remained a puzzle to me ever since. Why indeed did the naval authorities give their approval to my visiting their chief naval shipyard, and being shown their latest submarine engines only two or three days before the outbreak of hostilities?

I find it hard to recall my emotions at that time. I think perhaps the term 'bewilderment' most nearly sums them up. The newspapers did their best to comfort us by predicting that the war would be over before Christmas; that the Germans and Austrians could not possibly hold out for more than two or three months. There was the popular belief that our allies, the French, had the most powerful and best-equipped army in the world; that despite the recent activity in the German naval shipyards we still had by far the most powerful navy in existence with which effectively to blockade Germany; that our regular army, though small, was very well trained and was, as we were told, well equipped; that Russia, still smarting from her recent defeat by the Japanese, had built up a very efficient and well-disciplined army and had an almost unlimited reserve of manpower at her disposal and, finally, that we could almost certainly count on the support of Italy, with her powerful Armstrong-built navy, to control the Mediterranean. In short it was sheer madness on the part of Germany and her swashbuckling Kaiser to provoke a war with such doughty opponents. We were told both by our press and politicians to carry on with our business as usual and prepare to give a rousing Christmas welcome to our victorious army on its return.

As is well known, our army was shipped to France within a few days of the declaration of war. At once it advanced into Belgium in readiness to repel any invasion of that country by Germany and for the next two or three weeks the news was confused and conflicting. Then, in quick succession, came three shattering blows: to our dismay we learned that the British army, outnumbered and outgunned, had been forced to retreat from Mons, that the Russian army of nearly half a million picked troops, after invading East Prussia, had been surrounded and virtually annihilated by the Germans under von Hindenburg, and thirdly that the French had suffered a shattering defeat in the neighbourhood of Metz.

Realisation that the Germans had won these resounding victories almost simultaneously on both their eastern and western fronts, came as a terrible shock and at once put an end to all wishful thinking about the war being over before Christmas, or indeed that any end was in sight. Instead we were faced with the realisation that a long, grim and deadly struggle lay ahead of us, and that it was up to all of us in this country to play our part, either in the actual fighting or in support of the active services. The call went out for volunteers and for a time the recruiting offices were snowed under by the immediate response, but there were no adequate training facilities for the vast number of volunteers who offered their services. As a first step the number accepted was limited to 100,000.

In addition to our regular army, we had a fairly large reserve force of so-called Territorials. These were, for the most part, professional men from all walks of life who, in peace time, had volunteered to undergo some military training during their spare time. These were now called up and were being sent to France to replace members of the regular army who had borne the brunt

of the fighting during the retreat from Mons, and who were bei
sent home to train the thousands of new volunteer recruits. Mea
while a survey was made of the country's manpower in order
reserve certain individuals who, by virtue of their professional
technical knowledge, would be needed at home. As the on
mechanical engineer in my grandfather's firm I was classified
one of these and was instructed not to volunteer for active milita
service abroad. Many, if not most, of my Rugby and Cambrid
friends not so classified were killed during the ghastly struggle
the next four years.

By the late autumn of 1914 the first furious onslaught of t
German armies on both fronts had spent itself, and both sides ha
dug themselves into entrenched positions, facing each other acrc
a narrow strip of 'no man's land'. A period of trench warfare ha
begun and continued for the next four years with very little chan;
of position, both sides digging themselves ever deeper into tl
ground and fortifying themselves with innumerable reinforc
concrete machine-gun posts, and protecting themselves from a
tack by endless barricades of barbed wire. As to myself, the peric
of the next few months was one of sheer misery. My instructio;
were to carry on as usual with my work for the firm until call
upon for some other duty, but there was no work for me to d
Both my assistant draughtsmen who were Territorials and
whom I had become very attached had been called up for acti
service before the end of 1914. Palmer had been called upon
resume his former post as Chief Engineer to the Port of Londc
Authority, a very important and responsible position in wartim
and he paid only occasional fleeting visits to our London offi
which was rapidly becoming reduced to a mere skeleton sta
presided over by Tritton. Nobody seemed to want me and da
after day I sat alone in my empty drawing-office hoping that son
use would be found for my services. To while away the time I wei
on with the writing of my book and putting finishing touches
my design of an aeroplane engine, but with a heavy heart. Mo
of the time I spent reading reports from correspondents at tl
front which grew gloomier and gloomier every day, or scannin
the ever-lengthening columns of the casualty lists. It seemed tha
the war on land had reached a stalemate, and could only en

when the last British soldier had killed the last German. I conjured up pictures of our men standing knee-deep, or even waist-deep, in ice-cold water-logged trenches, or being tangled up in barbed wire in no man's land. The squalor of it all was one of the most appalling features.

At home in the evenings and at week-ends I carried on with the research into the phenomenon of combustion and detonation and with experiments on the single-cylinder version of my aero-engine to be. For how long this period of utter frustration continued I cannot recall; it seemed ages but, in actual fact, could not have been more than a few months.

By Christmas 1914 almost all my inner circle of friends including Hetherington, Thornycroft, Welsh, Sassoon, Cavendish Butler, Lucas, Erasmus Darwin and others had all enlisted as volunteers, but owing to the pressure on training establishments they were told to go home and wait until their services were called upon. Their lot was far worse than mine, for not only did they suffer the same helpless frustration but the prospect ahead of them was only the misery and horror of trench warfare from which I was secure.

In the event, however, Hetherington was classified as unfit for active service on account of his short-sightedness and was drafted to the newly formed Aeronautical Inspection Department; Thornycroft, because of his specialised knowledge of wireless telegraphy, was drafted to the Royal Naval Air Service, another new formation; Lucas, now a recognised authority on instrumentation, was drafted to Farnborough, but the others I have named were all drafted into the infantry and during the years that followed all, except Cavendish Butler, were killed. Butler alone, after twice being invalided home with serious wounds, eventually survived. By contrast with their fate I could but feel that I was a frightful shirker, but I buoyed myself up with the hope that I might yet have a chance to play a useful part in the war.

In the spring of 1915 I came across a man named Barrington, a very able young engineer who had served his time with Rolls-Royce and had been enlisted into a squadron of the R.N.A.S. under the command of a young engineer named Wilfred Briggs, R.N., to whom Barrington introduced me. Commander Briggs was a great enthusiast, a man of vision and imagination; the squadron he commanded devoted itself to the problem of anti-submarine warfare. He was a great believer in the possibility of long-range flying-boats which could both detect and attack enemy submarines and for this he contended that we should need a very much larger machine than anyone had yet contemplated, a machine that could range far out into the Atlantic and, at the same time, carry a worth-while load of depth-charges and a small high-velocity gun firing armour-piercing shells. Considering the state of the art in 1915 this was indeed an ambitious project, but Briggs was not to be deterred. He maintained that in his view a single large engine should be installed in the hull of the boat with twin propellors mounted on the wings and driven by means of chain or bevel gears much as in the original Wright machine. He said his advisers had told him that at least 600 h.p., and preferably more, would be required. At that early date the largest British-designed aero-engine, the R.A.F., was of only 140 h.p., while the French engines we were buying or starting to build under licence were all of still lower power, and their specific fuel consumption very high. I told him about the engine I had designed to operate with a stratified supercharge which could be employed either to boost the power or used as a diluent to reduce the fuel consumption under cruising conditions. He said that these were exactly the characteristics he was looking for. I told him that at best my engine would develop less than half the power he needed. He told me not to mind that but to get busy and design a larger version; after all, it only meant increasing all dimensions by about 30%! I told him that it was not quite so simple as that but that I would certainly look into what could be done. Briggs, like my cousin Ralph, was a go-getter and an optimist and not being himself very technical, relied on Barrington as second in command. He at once sent Barrington to examine my design of which, with certain reservations, he fully approved. Next day Barrington came to Walton-on-Thames to carry out tests in my workshop. My single-cylinder unit was in good fettle and put up an excellent performance, in fact it did a little better in power output and fuel consumption than I had claimed, and Barrington was much impressed. He was impressed, too, by the other work I was doing on the Dolphin engine on

THE FIRST WORLD WAR 153

the problems of combustion and detonation, the significance of which he fully understood. In particular he was fascinated by my Hopkinson optical indicator as an instrument for research.

A day or two later I had a visit from Briggs accompanied by three officers from his squadron and for their benefit I repeated the tests I had done for Barrington and explained and demonstrated the other work I was doing on the combustion side with the help of my Hopkinson indicator. They were all very interested in what I had to show them but were shocked at the condition of my aged five-inch lathe. I told them that it had originally belonged to an uncle of mine, that I had had it since the age of eleven, and it had served me well for more than twenty years but was now getting in a very shaky condition. They insisted that I must have something better. From some Admiralty store Briggs conjured up a brand-new six-inch precision lathe, a really beautiful tool, which arrived a few days later together with a new sensitive drill, both of which were Admiralty property and which I was allowed to keep until after the end of the war when we bought them in for our new establishment at Shoreham. A little later he offered me the services of a first-class mechanic named Doughty who had enlisted in his squadron and who proved invaluable to me. Up to this time, apart from occasional help by friends, I had been working single-handed and progress was necessarily slow but, from then on, with a full-time assistant and new up-to-date tools I was able to do far more.

I gather that Barrington must have given Briggs a glowing account of what he had seen in my workshop at Walton and of the design of my would-be aero-engine, for only a day or two later Briggs invited me to go with Barrington and himself to visit Messrs. Peter Brotherhood's works at Peterborough and to bring my designs with me.

Brotherhoods, at that time, were engaged on the design and manufacture of auxilliary machinery for the Navy, such as air compressors, steam pumps and such like. They were also in large-scale production of engines for torpedoes, further development of their famous little radial cylinder steam engine, so popular for contractors' use about the middle of the last century. From a steam engine it had been converted to run on compressed air, and had been widely used in mine working for the operation of coal-cutting machinery. By virtue of its compactness it had been adopted by the Admiralty for the propulsion of torpedoes for which purpose it was, and still is, admirably adapted.

As a naval engineering officer Briggs was well acquainted with the firm's products and had a profound admiration for its Chief Engineer and Managing Director, Commander Bryant, also an ex-navy man. The object of our visit was to find out whether Bryant would be prepared to develop and manufacture a much larger version of my design. In vain I had pleaded with Briggs to employ two or more smaller engines mounted in the wings but he had been insistent on the use of a large engine installed in an engine-room inside the hull and under supervision by a skilled engineer, as in naval practice.

Our visit was successful. Bryant agreed entirely with Briggs that a sort of flying destroyer was what the Navy needed and that it would require an engine power of at least five times that of any aero-engine so far developed by the British or French, and although a plunge deep into the unknown, it should not be ruled out on that account. He argued that 'nothing venture, nothing win' was the appropriate motto for wartime. He told us that although his firm was working twenty-four hours a day on the production of torpedo engines and other naval equipment, his large drawing-office and general purpose machine-shop were not so fully engaged and were in want of something interesting to get their teeth into. He then sent for one of his senior designing draughtsmen and his foundry foreman. We all settled down to a technical discussion from which it emerged that they would like to have a shot at designing and building a much larger version of my design, were ready to start work on it at once and that they would rush through in advance a single-cylinder version. It was further agreed that I should stay at Peterborough for a few days to supervise the layout of the new design and provide such data as I had gleaned from about two years' test work on my own small single-cylinder unit, and thereafter pay frequent visits to discuss details of design and supervise tests on their single-cylinder as soon as it was ready.

I was, of course, delighted; not only was it a job after my own heart but at long last it was one which I could feel justified my being reserved as a key man, and this took a great load off my conscience. At the time of this our first meeting, Bryant must have been in his middle forties. He was apparently in sole command of the very large firm of Peter Brotherhood which had recently transferred from London to a new and up-to-date factory at Peterborough. During the weeks that followed I came to know him intimately. He had an extraordinary grasp of every detail of the work in each department of his large works and obviously inspired the loyalty and affection of all ranks of his employees. I soon came to look upon him as a guide, philosopher and friend. At our first meeting in company with Briggs and Barrington I had confessed my misgivings about the use of a single engine and that of a size so far beyond my range of experience, or of any contemporary experience for that matter, but Briggs had been insistent that for long endurance in the air it was essential that the engine should be accessible to, and watched over by, an engineer with one hand on its pulse and the other wielding an oil-can and ministering to its needs. This, he said, was established naval practice. As to the rest of the machine Briggs said that he had been in touch with Messrs. Short of Belfast, who were prepared to design and build such a large flying-boat, a biplane with a stepped hydroplane hull, and that they, too, would prefer a single engine installed in the hull, if one of sufficient power could be made available.

During the next two or three months I spent several days a week in Brotherhood's drawing-office discussing the design of this much enlarged version of my aero-engine with their foundry and machine-shop managers, all of whom were most friendly and cooperative. During the previous three years I had drawn out several versions. That which I most favoured involved a single-piece aluminium casting embodying both the upper half of the crankcase and cylinder blocks, with wet liners and detachable cylinder heads, overhead camshafts and with four inclined valves to each cylinder. This looked feasible for the relatively small size of engine I had in view with a total cylinder capacity of about 15 litres, but not so when expanded to the 35 to 40 litres of our enlarged version. Their foundry expert, a very able and experienced man, doubted the possibility of producing so large an aluminium casting for the main carcase of the engine on account of the large contraction on solidification of this material; we agreed, therefore, to make it in two separate halves split vertically down the centre line of the engine, the two halves then to be bolted and dowelled together and thereafter to be treated as a single piece. He boggled also at the size and complexity of the long multi-valve cylinder head, and pleaded for separate heads to each cylinder, to which, with some reluctance, I agreed, for it involved increasing the cylinder centres and so adding several inches to the already long engine. As to the material of the cylinder heads, the choices were to use cast-iron, fabricated steel, or aluminium. Of these, fabricated steel was ruled out as too difficult and complicated; even with individual heads to each cylinder, cast-iron would be unduly heavy; there remained the third choice, that of using aluminium heads, but these would involve inserted steel or iron valve seats and no one at that date had had any experience of joining two materials with such widely different coefficients of thermal expansion. After much discussion it was agreed to try out an aluminium cylinder head on the single-cylinder unit and, if this failed, to fall back on the use of cast iron.

From these and such-like discussions with Brotherhoods I learned a great deal of value to me in the years that followed, but my misgivings deepened as to the success of our ambitious venture. These I conveyed to Commander Bryant who was always comforting and told me not to worry. He went on to say that, in his view, as an ex-naval engineer, the greatest danger we had to face was the submarine menace which had already accounted for many of our supply ships and which would increase in intensity as time went on. He said that although the Navy had done what they could to protect our fighting ships from torpedo attack, such measures did not apply to the vast number of unarmed and unprotected supply ships upon which our very existence depended. He thought that the only effective way of hunting down and destroying enemy submarines was a large flying-boat, such as Briggs envisaged, in that its high altitude would give it a wide range of vision, its speed would be at least three times that of the fastest naval surface vessel, it would be immune from torpedo attack and probably from any anti-aircraft gun-fire from an enemy submarine. Such efforts as we were embarking upon with this large engine would be justified fully if it saved the life of only one supply ship or caused the destruction of only one enemy submarine. If successful it might go far to rid us of this submarine menace; if it failed, well, it would pass into limbo like so many other ambitious projects, leaving us with the feeling that we had both done our best and gained a lot of valuable experience.

Like Briggs he had no faith in our small dirigible balloons or long rigid airships which Briggs had described as 'perambulating gasbags', whose speed was so low and range so short that they could not go far off shore or carry any offensive armaments. Thus comforted we agreed upon and completed the general design of our twelve-cylinder engine and set to work on the detail drawings, with the help of three or four first-rate draughtsmen. At the same time, work was started on a single-cylinder unit of $5\frac{3}{4}$ inch bore and $7\frac{1}{2}$ inch stroke. In what seemed to me an incredibly short time the engine was completed and ready for test. As a start we decided to employ an aluminium cylinder head embodying cast-in steel valve seat inserts for the two exhaust valves, while the two inlet valves and their seatings were in detachable cast-iron cages, as in gas engine practice. The crosshead piston and its guide were scaled-up replicas of those in use on my own small unit, the piston head and trunk being of 12% copper aluminium alloy, the lower end of the trunk being surrounded by a thin cast-iron sleeve, to act both as a bearing surface in the guide and to locate the floating gudgeon-pin. The compression ratio, as far as I can remember, was well under 5 to 1. After the usual preliminary testing and running-in, the engine initially behaved well in that with full supercharge it gave a brake mean pressure of just over 150 per square inch at a speed of 1800 r.p.m., but after a short run at this output, severe pre-ignition set in. On removing the head we found that both the exhaust valves were badly burned, and that the cast-in valve inserts were both loose. A new head was then made with screwed-in inserts. This behaved better and allowed us to run at full load for nearly an hour before pre-ignition set in. Again we found both exhaust valves badly burned and the valve seat inserts loose. We made several more attempts but, do what we would, we failed to make a secure job, so reluctantly we had to abandon the idea of using aluminium cylinder heads and fall back on the use of cast-iron. In those days neither low-expansion silicon aluminium alloys, nor high-expansion austenitic iron or steel was available, and perhaps we were attempting the impossible.

Apart from repeated failures due to burnt-out exhaust valves, the single-cylinder unit performed very well indeed both as to power with full supercharge and low fuel consumption, when using the supercharge air as a diluent. In particular the crosshead piston behaved perfectly in that it kept to all appearances very cool, and its lubrication was under complete control at all speeds and loads. While I was working on the single-cylinder unit, the drawing-office were getting on apace, patterns for the large crankcase were completed, and a sample casting of one of the two halves had been produced. This, at first sight, appeared quite sound but on cutting it up we found a number of blowholes and porous patches which would render it quite unfit for use. Our foundry expert seemed satisfied that with the judicious use of chills, by certain minor modifications to the design, and by more thorough venting, he would, in time, be able to produce a sound casting, but that it would require a lot of patient development work. I recalled my earlier childhood attempts at making castings with low melting-point alloys in plaster-of-paris moulds, and to the difficulties I had encountered when using alloys with a high coefficient of contraction so, though disappointed, I was not surprised at the failure of our first attempt.

Thus far we had got by the summer of 1915 when Bryant received a letter from the Admiralty instructing him to stop all further work on the flying-boat project. What exactly had happened I never knew. My impression is that Briggs had fallen foul of his superiors at the Admiralty, had been moved from his post of the command of a squadron in the R.N.A.S., and replaced by a more amenable naval officer who belonged to the lighter-than-air school. I could well imagine that so enthusiastic, impetuous and outspoken a man as Briggs could well be an embarrassment to his superiors, but all this is conjecture. A few weeks later I learned, to my great regret, that Briggs had died, I believe from pneumonia. Barrington, his staunch supporter and technical adviser, had transferred to another squadron and I lost touch with him for the time being.

Thus ended my association with the flying-boat project and for the time being with Messrs. Peter Brotherhood. During the four months it lasted I became devoted to Commander Bryant and he, I think, took a fatherly interest in me and my future. When at Peterborough I always stayed at his house where he and his wife treated me as one of the family. I was, I suppose, a good listener and I loved listening to his accounts of his early life in the navy and of his early struggles after he left the sea to go into industry, and of the many mistakes and failures he had made in the course of his climb to his present position as technical head of a large industrial firm.

After the winding-up of the flying-boat project I returned once again to my lonely office in London where I found conditions had improved in that the firm was being consulted on a number of problems connected with the British railways and those in Northern France temporarily under British control, but even so there was very little for me to do. My four months of activity and Bryant's encouragement had done much to restore my spirits and to revive the hope that I might yet have a chance of playing a useful part in the war effort. In the meantime I set to work to finish the first volume of my book. This I completed and sent off to Blackie. They expressed themselves as pleased with what I had done but regretted that owing to shortage of staff and other difficulties they would not be able to publish it until after the end of the war. Very generously, however, they sent me a cheque for £250 due on completion of the manuscript of both volumes, although I had not yet made even a start on the second volume.

Although the flying boat project had been wound up, I was allowed to keep both my new lathe and the full-time assistance of Doughty who, in fact, was to remain with me for the next fifteen years. Doughty before the war had run a small shop as a cycle repairer in Walton-on-Thames. He had volunteered and, being an expert mechanic, had been drafted into Briggs' squadron of the R.N.A.S. He was indeed a very skilled mechanic; he knew it, and was proud of it but, in many ways, he was a difficult man for he had a violent temper which flared up at any hint of criticism, and he was no respecter of persons. It took me some time to learn how to handle him, but when he realised that I, too, was a good mechanic, we got on splendidly and I could not have wished for a more loyal and capable assistant. I suspect that the reason the R.N.A.S. allowed me to retain his services throughout the war period may well have been that he was too much of a stormy petrel and too intolerant of discipline for their liking.

As I have said earlier in this chapter, my friend and colleague, Harry Hetherington, had joined the Aeronautical Inspection Department, the head of which was General Bagnall-Wilde. Hetherington had been allotted the task of analysing and reporting on captured aero-engines, a task after his own heart and for which he was ideally suited. He had told Bagnall-Wilde about the 300 h.p. supercharged engine I had designed which, by virtue of its supercharge, could be capable of attaining a very high altitude. At that time we were suffering from bombing attacks carried out by the gigantic Zeppelin rigid airships which flew at an altitude of between 16,000 feet and 20,000 feet and well out of reach both of our aeroplanes and our few anti-aircraft guns; thus the need was for an aeroplane which could climb up and attack these monster airships. Bagnall-Wilde was not himself very technical, but he had on his staff a very capable young flying officer, a certain Captain Halford, whom he sent to Walton to examine and report on my proposed design and to witness tests on my single-cylinder unit. Halford, like Briggs, was a great enthusiast; himself a pilot he combined a shrewd judgement with a sound knowledge of theory and practice. At once he grasped what I was getting at and agreed that in the existing state of the art, it should be possible, with 40% supercharge, to reach an altitude of well over 20,000 feet. He must have reported very favourably on his visit for Bagnall-Wilde instructed him to keep in touch with me and, through his experience as an inspector, to find a firm competent to carry out the manufacture and development of such an engine. Halford chose the firm of Beardmore, with whom he was well acquainted. Beardmore had been building the 120 h.p. Austro-Daimler engine, a licence for which they had taken out before the war, and they were also at that time developing a six-cylinder, 250 h.p. engine, largely of Halford's own design. This engine, to be known as the 'B.H.P.' was already in an advanced state of

Triumph Ricardo Motor Cycle, c. 1922

Dolphin Engine, c. 1907

Mk. V Tanks. Battle of St. Quentin, 29th September 1918
By permission of the Imperial War Museum

*Medium Mk. 'B' Tank at Metropolitan Carriage Wagon & Finance Company's
Works. (Left to right) Maj. J. Buddicom, Lt. Shaw, R.N.V.R., H.R. Ricardo,
Maj. W. G. Wilson*
By permission of the Imperial War Museum

development. Several prototypes had been completed and were
undergoing tests prior to putting it into production. Halford had
a high opinion of Mr. Pullinger, Beardmore's Chief Engineer and
Technical Director ('B.H.P.' stood for Beardmore, Halford and
Pullinger). With him he had discussed my proposed design of
supercharged engine and had found Pullinger receptive. Next,
Halford brought with him to Walton Beardmore's Managing
Director, a rather taciturn fellow, and Pullinger, who seemed to me
a live wire indeed. Together we went through my drawings and ran
tests on my single-cylinder unit. Pullinger seemed much impressed
by the latter's performance and liked the general layout of my
design but he, like Brotherhood's foundry expert, had a horror of
large aluminium castings and urged that I should be content with
the use of separate cylinder blocks in place of my monolithic construction. This, of course, would add a little weight but would
simplify the foundry work and machining. It would mean, however, re-drawing the whole thing. After a full and useful discussion
it was agreed that, subject to Bagnall-Wilde's approval, Beardmores would undertake the development and manufacture as soon
as they had put the finishing touches to their B.H.P. engine. In
the meantime, Halford offered me the services of a competent
young draughtsman, Plunkett, who had been drafted into the
A.I.D.

Bagnall-Wilde gave his approval and Plunkett joined me in my
London drawing-office, where we set to work to revise the design
to embody separate cylinder blocks and to incorporate a few other
minor modifications which Pullinger had recommended. Plunkett proved an excellent assistant; though not very experienced
he was quick at understanding what was wanted, was conscientious and hard-working and a delightful companion. Incidentally,
Plunkett re-joined me after the end of the First World War and
remained with me until he retired in 1963.

All this, as far as I can remember, was during the winter of
1915–16; during the next few months Halford divided his time
between tours of inspection and alternate visits to me and to Beardmores at Dumfries. He was getting very fretted by the delays in
getting his B.H.P. engine finalised and into production, delays
caused largely by the changing requirements of active war service.

It did not take Plunkett long to complete the revised design of my engine which was sent to Beardmores for detailing in their drawing-office. This done, Plunkett was recalled to the A.I.D. and I was once more left alone in my empty drawing-office, for there was nothing more I could do till Beardmores were ready to make a start. Week followed week and nothing happened, and the news from the front grew gloomier and gloomier. In 1915 the list of killed filled an ever-lengthening column of *The Times*; by the spring of 1916 it no longer filled a column but whole pages, and showed that the death rate among our troops was averaging 1000 a day. Of my two pre-war draughtsmen one had been reported killed and the other missing. Next came the news that my friend, Tony Welsh, had been killed within a week of his arrival at the front, and so the ghastly slaughter went on and on with no end in sight; at both Charing Cross and Victoria Stations a seemingly endless succession of hospital trains unloaded the wounded into mile-long queues of ambulances. I well remember a Divisional Commander telling me that to keep his infantry division up to strength he had to budget on the assumption that the expectation of life of a junior officer was only three weeks in the front line : and here was I, sitting alone in my empty office, unable to help, waiting and hoping for some opportunity to turn up. In the event I did not have to wait much longer.

Before relating the immediate events I will describe what eventually was to happen to the supercharged engine. Before the outbreak of war I had become acquainted with a certain Mr. Edwin Orde who had joined the Board of Armstrongs. At the outbreak of war Orde had persuaded Armstrongs to establish and equip a department of light engineering in which to manufacture component parts and to carry out sub-contracting work for the various firms engaged upon the production of aero-engines, gun fittings and other small parts requiring a high standard of precision. In charge of this department were Le Mesurier and Everett, both able young men and under their direction the new establishment proved a great success. I had come to know Orde very well and in years gone by had told him about my design for a supercharged aero-engine and of my failure to get the War Office interested in it, or in aero-engines of any kind for that matter. Since the outbreak of war, however, the attitude of the War Office had changed completely and production of aero-engines had become a top priority. At some time in the winter of 1915–16 Orde had asked me what was being done about my aero-engine design. I told him about the flying-boat episode which had come to nothing, and that Halford and I had handed my design over to Beardmores and they had not yet made a start on it. He told me that Le Mesurier and Everett, who having by then made nearly all the individual parts of aero-engines for other firms, considered they had gained enough experience to build complete aero-engines on their own and that he – Orde – had told them about my design which had made a great appeal to them, all the more so because the crying need at that time was for high altitude performance. There, for the time being, the matter ended and the weeks dragged on and still Beardmores were not ready to make a start.

In the spring of 1916 I became involved with other matters and had to leave Halford to wrestle with Beardmores. In the months that followed it became evident that something was wrong with the firm of Beardmore. There had been violent upheavals in the Directorate of the firm, as a result of which Pullinger had left them and the whole establishment fell into chaos. What exactly happened I never knew, but the upshot of it was that to Halford's disgust the final development and production of his beloved B.H.P. engine was handed over to Siddeleys, and its name changed to 'Puma'. As to my engine this was once again put on the shelf. All this was, of course, a bitter disappointment to Halford and myself for it looked as if my engine, the design of which was by then four years old, would never materialise.

After the virtual break up of Beardmores, Halford got in touch with Orde and Le Mesurier and a contract was placed with Armstrongs for the construction of six prototype engines of my design, the engine to be known as the 'R.H.A.', the initial letters standing for 'Ricardo, Halford, Armstrong', but this was not until the spring of 1917. Armstrongs got busy at once and a beautiful job they turned out, and that in a remarkably short time, but it was just too late; by the date of the Armistice two engines had been completed and tested and the remaining four were almost complete. On its Type Test the first engine developed 260 h.p. without any supercharge, 300 h.p. when using the supercharge air as a diluent, and 360 h.p. when making full use of the supercharge air at ground level. The first of these engines had been sent to Farnborough for flight tests just before the Armistice in 1918 and was flown by Norman in a De Havilland plane, who reported that it behaved well and responded to supercharge as we had hoped. However, the war was over and, like so many other promising developments, the engine passed into the shadows.

Frank Bernard Halford
By permission of the RAeS

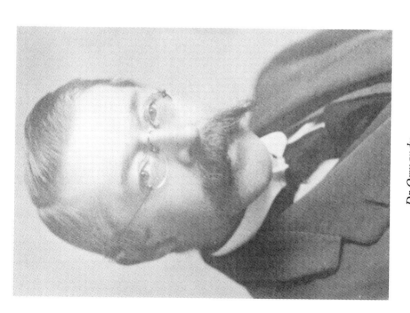

Dr Ormandy
By permission of the Inst. Mech. Eng.

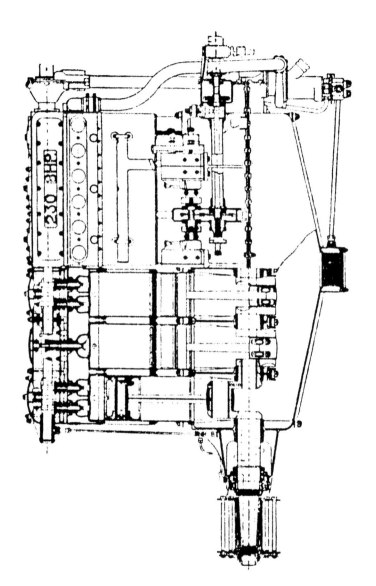

The B.H.P. engine, from which the Armstrong-Siddeley Puma was developed

CHAPTER 11

The First Tanks

By the spring of 1916 I was feeling really desperate; the war had already gone on for nearly two years and was growing more horrible every day while I, though in a safe and privileged position, had achieved nothing to help the war effort. The flying-boat project had got nowhere and week after week Beardmores failed to get started on my aero-engine.

By the late spring of 1916, when I had almost plumbed the depths of gloom, Tritton invited me to his office to meet a new client. He introduced me to a Lieutenant Francis Shaw, a youngish man in naval uniform and a member of Squadron 20 of the R.N.A.S. Tritton told me that Squadron 20 would like our help and advice about a certain problem which he thought was right up my street; that it was very secret and not to be discussed in our office, but that I was to go with Lieutenant Shaw to the Headquarters of Squadron 20, then an office in Pall Mall, where I would be told about it. There, for the first time, I heard about what later came to be known as 'Tanks', a development which had been carried out in great secrecy by the technical staff of that squadron. I was shown drawings and photographs of their latest Mark IV Tank. The problem on which they wanted my advice was how to get this huge machine, weighing 28 tons, into position on a specially-designed railway-truck. The overall width of the Tank was only an inch or two less than the railway loading gauge. It was essential, therefore, that its position on the truck should be accurately located throughout its entire length. It was easy, they said, to drive the machine up a ramp on to the floor of the truck under its own power, even though the steering was somewhat erratic. The problem was how, having got it there, to manœuvre or wriggle it into position. Shaw had heard, it seemed, of my design of jacks with hydraulic intensifiers which were being successfully used for manœuvring the span of a bridge into position on its pier, and wondered whether a similar device could be employed for their purpose. That was my first introduction to the Tanks which were to have such an important place in my thoughts for the rest of the war. At this, my first contact with Squadron 20, I was introduced to their Technical Committee.

I must apparently have made quite a good impression for in the weeks that followed I was invited to attend several more technical meetings when many other mechanical problems were discussed, and was taken to see the vehicles under construction and on test both at Lincoln and at Wolverhampton. In short, I seemed to have insinuated myself into the technical affairs of the enterprise generally, and was co-opted as an unofficial member of the Technical Committee. The initial problem on which my advice had been sought faded out of the picture as the drivers of the Tanks became more skilled, but there were always many other mechanical problems on some of which I was able to give a little help.

I think it may be of interest if I go back to the beginning and give my recollections of how the Tanks first came into being and how it was intended that they should be operated. After the first onrush of the Germans in the autumn of 1914 had spent itself, the fighting front became static and, in all probability, would have continued thus for both sides were deeply entrenched with second and third supporting lines. Between the front-line trenches lay a strip of no man's land pitted and churned up by gun-fire and covered by a jungle of tangled barbed wire. Any attacking force had to advance through this as best it could through heavy shell and machine-gun fire from numerous small reinforced concrete emplacements. Only by dint of an enormous concentration of men and guns on some narrow sector of the front was it possible to reach the enemy front-line trenches and the preliminary bombardment deemed necessary to cut the barbed wire gave the enemy ample warning of the time and place of the attack. If the preliminary bombardment had been really effective and the concentration of infantry sufficiently large, it was sometimes possible to capture a few hundred yards of the German front-line trenches, only to be driven out again by an enemy counter-attack. Such operations entailed an enormous loss of life with very little gain.

It seemed that unless some entirely new approach to the problem could be provided, only sheer exhaustion could bring the ghastly struggle to a close.

In 1914, at the instigation I believe of Winston Churchill, and certainly under his patronage, a formation known as the Royal Naval Air Service was created. It was made up almost entirely of civilian technical volunteers, and was divided into a number of more or less independent squadrons each commanded by a naval engineer officer. Each squadron devoted itself to some particular branch of the Service, e.g. one squadron commanded by Wing-Commander Cave-Browne-Cave, R.N., concentrated on the development of the small airships, observation balloons and kites, and another, under Commander Briggs, R.N., on anti-submarine devices, such as depth charges and so on.

Squadron 20 devoted itself to the development of an armoured vehicle capable of crossing wide trenches and of trampling down or tearing away barbed wire. In this they had the support of Winston Churchill and Lloyd George, but not of the Army authorities who thought the project merely silly.

A small design committee under the Chairmanship of Sir Eustace Tennyson D'Eyncourt, then the Director of Naval Construction, set to work on the development of H.M. Landships as they were then officially known. Of this committee the moving spirit was Sir Albert Stern, a wealthy banker who provided finance when needed. He was a capable organiser and a tower of strength and common sense. Perhaps the leading technical brain of the committee was that of Wilson, already well known as a great authority on gear problems. The construction of prototype machines was put in the hands of Messrs. Foster of Lincoln and their Chief Engineer, Sir William Tritton, who himself contributed some very able service and help. In what now seems an incredibly short time, a fourth prototype machine was evolved which could cross any trench either by spanning it or, if the width was too great, by falling into it and digging its way out again. It could cope with barbed wire either by simply treading it down or, when required, could sweep it away by towing grappling irons.

Even so, the Army would have nothing to do with the vehicle

although a few individual officers were quite enthusiastic. Part of the trouble may have been due to the fact that the whole development was carried out by naval personnel and jealousy may have played a part. Thanks to Stern's persistence and Lloyd George's help, and after a great deal of experimental work, a batch of fifty landships was built, partly by Fosters and partly by the Metropolitan Carriage and Wagon Works, Birmingham. All this had been carried out in great secrecy, but further tests could not be carried out without the appearance of the vehicle in public, and some explanation had to be offered for their strange appearance. It was accordingly put about that they were mobile watertanks for use by the troops. This appeared plausible and the name 'Tanks' came into being.

The plan of attack envisaged by the sponsors of the Tanks was as follows. On the sector selected for attack, the Tanks would be brought over during the night as close up to the front as possible, and camouflaged with netting or branches of trees or other methods. There would be no preliminary bombardment, nor would the enemy have any warning that an attack was impending. At zero hour the Tanks would cross our trenches and proceed across no man's land, each Tank being followed by as many infantry as could find shelter behind its armoured shield. Each Tank was armed with two six-pounder guns, one on either side in sponsons; thus they could knock out any machine-gun posts encountered on their way. Each Tank would leave in its wake a relatively clear path for further infantry to follow. Thus a powerful force could be landed by surprise in the enemy's front line trenches. The Tanks could then proceed to deal with the enemy's supporting trenches in the same way until a complete break-through had been achieved. This, at least, was the modus operandi envisaged by Squadron 20.

I was invited to attend a demonstration of the new Tanks which was being held on a test ground close to Wolverhampton and which was attended also by various 'high-up' officers in the War Office. A piece of waste land had been prepared to resemble as far as possible the typical fighting front in France, which comprised trenches of various widths and depths, deep areas of barbed wire and other such-like obstacles. I was deeply impressed by what I

saw of the ability of the Tanks to cross wide trenches, to dig themselves out of deep shell holes and to deal with the barbed wire.
The military authorities also appeared to be impressed, but they
maintained that the only test that would really convince them was
a test carried out against the enemy in France, and they insisted
that a certain number of the Tanks should be despatched over-
seas as quickly as possible. Accordingly about twenty or thirty
of the Tanks were manned by crews recruited from Squadron
20 few, if any, of whom had had any military training, and
despatched to France in September 1916. Almost immediately
they were put into action on a sector of the front which was
then comparatively quiet, and not very heavily manned by either
side.

The whole operation was a tragic example of mis-management
and muddle. Without any preliminary rehearsals, and without any
clear plans for co-operation with the infantry who apparently
were as ignorant of their purpose as the enemy, the Tanks were
instructed to cross no man's land, knocking out any machine-gun
posts on the way, to get astride the enemy's front-line trench, and
then to carry on as seemed best to them. The Tanks had no difficulty whatever in crossing no man's land, and after knocking out
a number of machine-gun posts on the way they arrived successfully astride the enemy's front-line trench, only to find, to their
dismay, that no infantry were following them, and that they were
left alone to their own devices. Most of them, having no idea
what to do next, returned to their base, but several carried on
straight ahead, crossing the enemy's second and third lines of en-
trenchment, and so out into the open country behind the enemy's
lines, only to run out of petrol. They did what they could to disable
their machines and then abandoned them and attempted to escape
back into our lines. Most of the crews were captured but a few
did succeed during the following night in getting back to our side.
The whole affair was really a disaster of the first magnitude for
the Germans now knew what to expect, and had ample time to
prepare counter measures in the form of light anti-tank guns, rifles
with armour-piercing bullets, landmines and other obstacles, while
we on our side had little or no hope of producing any effective
number of Tanks during the next twelve months. This at least

is the unhappy story brought home by the survivors. It may have been highly coloured for they were feeling bitter and frustrated, but it could not have been far from the truth. On the credit side, the military authorities could not ignore the fact that a handful of men, and a very small number of Tanks, had been able to break clean through the German defensive lines and, on the strength of this achievement, orders were given that the production of Tanks should proceed on a large scale. On the credit side, also, the returning Tank crews brought home much experience of value. For example, they reported that the store of ammunition carried was not nearly large enough, that larger fuel tanks were needed, and so on.

The tests at Wolverhampton had revealed, among other things, that the engines then used were not at all suitable. For the first batch of experimental Tanks, the 105 h.p. Daimler sleeve-valve engine had been employed being, at the time the highest-powered petrol engine for vehicles in full production. Its power output was barely adequate to deal with the adverse conditions when the tracks were cloggy with glutinous clay or entangled with barbed wire. Its lubrication system consisted of troughs into which the connecting-rod big ends dipped but it provided little oil cooling of the bearings and bearing failures were frequent. Again, the trough systems failed completely when the engine was tilted through any considerable angle in the fore and aft direction, as was the case when climbing out of a trench. With purely reciprocating sleeves there was nothing to spread the oil circumferentially, and seizures could be prevented only by supplying a very large quantity of oil to the outer sleeve. This meant that oil was carried into the exhaust port, which resulted in a smoky exhaust at all times and in the emission of clouds of blue smoke when the engine was accelerated after any prolonged period of idling.

It was relatively easy to screen the Tanks from aerial observation even when close behind the front line but all camouflage would be useless in the face of exhaust smoke. It seemed that all the automobile manufacturers already had their hands full either with military vehicles or with aircraft engines which, by that time, had sprung into first priority. It was apparent, therefore, that an entirely new engine would have to be designed and built. As I have related in Chapter 8 I was acquainted with a number of the manufacturers of large industrial engines such as Mirrlees, Crossley and others and I was asked to canvass these to find out whether they had the capacity and were willing to tackle the construction of such an engine, for after we had handed over the samples of our Tanks to the Germans, there was no longer any need for secrecy. I found them not only willing but eager to do this, but not to tackle the design which they said was quite outside their range of experience. I reported this to Tennyson D'Eyncourt who suggested that I seemed to have a good idea of what was required and asked me to undertake it. I told him that although I had always been deeply interested in engine design, my experience was really very limited, but I should love to try provided that I could be assured of the goodwill of the would-be manufacturers, but that I was afraid that the engineers of such firms, with a hundred times my experience, would regard it as an impertinence to be asked to build an engine designed by a little-known young man.

I had, of course, already given a great deal of thought to this problem of an engine for the Tanks, and had formulated fairly clear ideas of my own as to how it should be designed. It was left that I should go home and prepare notes of my views, together with a rough scheme design of the kind of engine I had in mind. This done, D'Eyncourt would call a meeting of the would-be constructors at which he himself would take the chair, and we should discuss the design and my notes. If the reaction was favourable, then we should go ahead with all speed.

The meeting was a great success. Sir Eustace handled it admirably. My notes and my proposed design had already been circulated and were very well received, in fact all present were very flattering about it. Sir Eustace impressed upon them the need for urgency and pointed out that I would be in need of and would welcome all the help and advice they could give me. One and all promised that they would co-operate whole-heartedly both with me and with each other in the development of the new engine. At this meeting it was agreed that Mr. Windeler of Mirrlees would act as co-ordinator among the whole group, and that the firm of Mirrlees would be the parent firm. The group would consist of

Messrs. Mirrlees, Bickerton & Day, Messrs. Peter Brotherhood, Messrs. Browett-Lindley, Messrs. Crossley Bros. and Messrs. Hornsby. To these were later added Messrs. Gardner and the National Gas Engine Co., making seven in all. It was also arranged that Mirrlees should lend me two junior draughtsmen, A. Ferguson and G. Holt, and that the scheme design, with their assistance, should be carried out in the offices of my grandfather's firm, while the detail working drawings would be prepared by Messrs. Mirrlees under Mr. Windeler's direction. It was agreed, also, that the power to aim for was approximately 150 h.p. I urged for rather larger power, something over 200 h.p., but Wilson, who had been responsible for the design of the transmission system of the existing Tanks was doubtful whether it could be made to cope with more than 150 h.p. Never in my life have I attended a meeting at which so much was settled in so short a time, for I came away with a mandate to go ahead with the design.

I ought perhaps to explain that I was already well acquainted with the personnel of these various companies and that politics played an important part in the choice of firms. The R.N.A.S. and Squadron 20 were naval units and all the firms in question were, at that time, engaged on naval equipment, such as submarine engines and auxiliary engines for naval purposes of all kinds. All of them were heavily committed, but all said that they could each make a relatively small number of engines for the Tanks. At that time no one had any definite idea as to the numbers that might be required, but the general guess was that it would be of the order of two or three hundred. In fact well over eight thousand of the engines were built before the end of the war.

The experiments carried out at Wolverhampton, coupled with the brief but disastrous adventure in France, had brought to light a great many features which needed to be redesigned. The Mark IV Tanks relied on clutch and brake steering, that is to say, a clutch was provided for each track and also a brake. It called for rather extreme muscular effort to operate these clutches and brakes, and together with the speed control and gear changing was too much for a single driver. In practice it really took three men to drive the Tank, a driver in the centre to control the engine and change gear, and one man on each side track. This, of course, was an extremely cumbersome arrangement and only possible by virtue of the fact that the speed of the Tank was restricted to that at which the infantry could follow comfortably on foot.

Wilson had meanwhile been working out and developing his epicyclic gear whereby the torque could be shifted from one track to the other without any appreciable effort on the driver's part. Thus, at the beginning of October, 1916, we started on a complete new design : a new design of hull, a new design of transmission by Wilson, and a new engine. Meanwhile, to fill the gap before these could be ready it was decided to carry on with the original design known as the Mark IV, for it was thought that it would be nearly a year before the new Mark V Tank would come into production.

With the knowledge that several of the early Tanks had already fallen into enemy hands, neither the vehicle itself nor its potentialities could any longer be a matter of surprise to the Germans, and it was obvious that the winter of 1916–17 would be devoted by them to the preparation of counter measures, as indeed proved to be the case. There was, however, still the element of tactical surprise, for possession of the Tanks made it possible to attack on any sector without any preliminary warning in the way of massive artillery preparation, but to achieve tactical surprise it was essential that the vehicles should not advertise their position by the emission of clouds of exhaust smoke.

In those days there were no such things as oil control rings for pistons, and blue smoke in the exhaust was a prevalent nuisance. Some three years previously I had designed and built myself in my small workshop an experimental single-cylinder four-cycle engine in which I used a crosshead-type piston employing the lower side of the piston crown as a supercharger. The crown of the piston was thus isolated from the crank-chamber and very effective oil control could be maintained and the exhaust was completely invisible. Apart from the question of oil smoke the crosshead piston had many other alvantages less obvious perhaps at first sight. I therefore decided in the new design to adopt this type of piston, but not to use it as a supercharger. Instead, the air on its way to the carburettors was circulated around the crosshead guides and under the piston crown. This raised the temperature of the air entering the carburettors just enough to deal with the carburation of the very poor volatility fuel that was available at that time. By this means also heat from the piston crown was not imparted to the oil, and the oil in the crankcase kept cool without the need for any external oil cooler; also the oil consumption was extremely low. It was intended at the time that overhauls to the engine should be carried out in position in the Tank, and since it was impossible to get underneath the engine, this meant that the crankshaft must be carried in the bedplate and accessible from above, as in marine or large stationary engines. As to the cylinders, combustion chambers and valve layout, I followed exactly the arrangement I had used in my single-cylinder engine. This was quite conventional, and had given a good performance, although it was very prone to detonate. I therefore employed a compression ratio of only 4.3 to 1, and even this low ratio appeared to be rather on the high side for the very poor fuel with which the Tanks were supplied. When climbing out of a wide trench the Tanks often had to operate up to an angle of 35 degrees, so that the lubrication system had to cater for this.

I was naturally worried about the dangers of torsional vibration in a long six-cylinder engine and, to be on the safe side, I fitted a Lanchester torsional damper of the viscous type. We had, however, great trouble in retaining the oil in the dampers. Gardners substituted a damper of their own design. This was a single disc faced with Ferodo and running dry; it was completely successful and gave no trouble, but it had one serious besetting vice and that was that when in action it was liable to squeak loudly, to the alarm of the driver. We never found a remedy for this.

I have said earlier that Lloyd George was a firm supporter of the Tank project. Early in 1916, I think, he had become Minister of Munitions with very wide powers and he was also a personal friend of Sir Albert Stern whom I regarded as the most active driving force behind the whole project. The War Office, however, was still only lukewarm, though a few individual members of the Staff, in particular Major Ellis and Colonel Swinton, were staunch supporters of the project. At that time, and indeed throughout the earlier part of the First World War, I was still in the employ of my grandfather's firm and on their pay roll, but was loaned to the Ministry of Munitions and later to the Air Ministry, but with no clearly-defined position with regard to either – a sort of roving commission. This was admirable from my point of view. As a civilian I was free to criticise or even abuse the military authorities without risk of court-martial, nor could I be sacked from a position which officially did not exist.

Lloyd George had instructed Squadron 20 to push on with all possible speed with the design and development of the new Mark V Tank and had promised that his department would handle the financial and contract side of the work. With the help of Ferguson and Holt the design work proceeded rapidly and smoothly. I could not have hoped for two more able and helpful assistants. Ferguson stayed in London with me while Holt divided his time between my office in London and Mirrlees in Manchester, where the detail working drawings were being prepared.

During this time I had many meetings with the various manufacturers at which we discussed all the technical details. At these meetings Windeler usually took the chair with great tact and charm of manner. He was regarded as the undisputed leader of the group and was held in great respect by all the parties.

It was a little later that the firm of L. Gardner & Sons joined the group and I was a little apprehensive about this at first, for Gardners had had a vast amount of experience in the building of high-speed engines which the other members of the group lacked. I was afraid, therefore, that they might be very critical of my design and push for a number of changes. My fears, however, proved entirely groundless. They accepted my design as it stood and I found them co-operative and altogether quite delightful people to deal with.

All but one of the group were in the Manchester area and in close touch with one another. The single exception was the firm of Peter Brotherhood in Peterborough. I was very anxious to include Brotherhoods because, as I have already explained, I had a profound admiration for Commander Bryant, their Chief Engineer and Managing Director. He realised the responsibility I was undertaking and was determined to make things easy for me in every way he possibly could. He liked my design but pointed out that being decidedly unorthodox there were bound to be teething troubles. He therefore offered to rush one engine through as quickly as he could and let me have it to play with. This was a very generous offer which I accepted gladly.

While the engine design was in progress Wilson was hard at work on the transmission gear for the new Tank. He, too, had collected together a group of firms to produce his gears. Wilson at that time was a serving member in Squadron 20, and since their headquarters were in London, he and I kept in very close touch. At the same time Tritton was busy with the design of the hull and he, too, had collected together a group of firms, mostly locomotive or rolling-stock manufacturers, chief of which was the Metropolitan Carriage and Wagon Works.

Just before Christmas of 1916 Lloyd George informed us that he was placing orders for 1400 of the new Mk. V Tank, no single part of which had yet got further than the drawing-board. This was rather alarming to all of us but Commander Bryant was as good as his word and had an engine running on the test-bed by the beginning of March 1917. This was a great achievement for it gave me a clear six weeks in advance of the other members of the group to play with the engine and get the bugs out as far as possible. Thanks to the very generous help and advice I had received all through, not only from the manufacturing group but from other friends in the aero-engine and motor-car industries, such as Rowledge of Napiers and Pomeroy of Vauxhalls, and partly also, I think by sheer good luck, the first engine performed quite well from the outset and no very alarming symptoms developed. Teething troubles there were, of course, but, for the most part, they were pretty obvious and easy to overcome for, at Brotherhoods, I had the assistance of an experienced foreman tester. If I wanted a new part, or a part modified, it was done during the night and available for me next morning.

By the end of April Mirrlees had their first engine running and a beautiful job they made of it. Almost straight away it exceeded its rated power of 150 h.p., for it developed, I remember, 168 h.p. at its governed speed of 1200 r.p.m. and just over 200 h.p. at 1600 r.p.m. It was exhibited to, and put through its paces, before the other members of the group. Crossleys were next, I think, to get an engine running a week or two later. It had been agreed that each firm in the group would keep at least one engine continuously on the test-bed so that experiments and modifications could be tried out.

By the summer of 1917 the group was turning out about forty engines per week. We were well ahead of both Wilson and Tritton. Meanwhile our engines were being used to replace the Daimler engines in the Mark IV Tanks. We had six engines set aside on permanent test-beds for endurance running and for testing modifications and these proved a great boon. Each production engine had to undergo a one-hour acceptance test at 150 h.p. at 1200 r.p.m., followed by ten minutes with a wide open throttle at 1600 r.p.m. This was the routine acceptance test but, from time to time, one engine was picked at random and submitted to a ten-hour full-load test and to a very thorough calibration.

The first Mark V Tank was completed, I think, early in June, and one of the new engines was duly installed. On the whole it behaved very well and Wilson's new epicyclic gear was a great success, so much so that the Tank could be steered quite comfortably by a single driver with very little muscular effort. One defect of the new engine was that there was far too much exhaust pipe inside the hull and the heat became intolerable. In heavy going the exhaust manifolds glowed a bright red heat. To remedy this a sheet-metal cowling was fitted around the exhaust manifold, through which air was circulated by means of a belt-driven fan, and discharged out through the roof of the Tank. This was a makeshift arrangement but it served its purpose.

These relatively large engines had to be started by hand, which was no idle task. The danger of the engine stalling, when in operation, had been impressed upon me, for a stationary Tank in the middle of no man's land would be a sitting target for the enemy's artillery. The Daimler engine, having a very light flywheel, and a not very satisfactory clutch, was liable to stall when changing gear. To reduce this risk I had provided what we called an anti-stalling governor, that is to say a governor whose function it was to open the throttle when the engine speed fell below 400 r.p.m., similar to the idling governor which we use on diesel engines today. This had appeared to work very well on the test-bed but, in actual service, it was an utter failure, for it tended to open the throttle far too rapidly and thereby stall the engine. It was therefore abandoned. Wilson had designed a new gearbox and clutch consisting of a series of flat plate clutches lined with Ferodo and the combination of this really good clutch and a heavy flywheel provided an acceptable safeguard against the risk of accidental stalling.

The production of engines for both Mark IV and Mark V Tanks kept well ahead of that of the hulls and by the end of 1917 the 150 h.p. Tank engine was being delivered at the rate of one hundred a week. By this time the military authorities had been completely won over. A new branch, the Royal Tank Corps, had been formed, and the demand was for more and ever more Tanks. Meanwhile the enemy had developed, among other devices, a heavy rifle fired from a tripod with armour-piercing bullets which could penetrate the armour of our Tanks. To counter this new weapon the thickness of the armour-plating had to be increased. This added several tons to the weight of the Tank.

It seemed fairly obvious that as time went on the Tank would become heavier and larger, and the weight of the armament would increase. I was therefore asked to undertake the design of a larger engine, a six-cylinder of about 50% larger cylinder capacity. As soon, therefore, as the design of the 150 h.p. engine had been completed, and sufficient tests carried out to satisfy us all, I started on the design of the larger six-cylinder engine. In this the piston diameter was increased from $5\frac{5}{8}$ inches to $6\frac{3}{4}$ inches but the stroke of $7\frac{1}{2}$ inches remained the same. In the new design, which was known as the 225 h.p. engine, I employed four horizontal valves operated by push-rods and bell cranks from two camshafts. In the case of the 150 h.p. engine I had taken a great risk in adopting a crosshead piston, but I did not feel justified then in risking any other departure from orthodox practice or from that with which I had had personal experience on my own home-made engine. The cylinders, the disposition of the valves, and the whole of the valve mechanism was therefore a scaled-up version of my smaller engine, as was also the form of the combustion chamber. Apart from its proneness to detonate, this form of chamber and valve layout compelled the use of exhaust manifolds with several changes of direction inside the body of the Tank. In the new and larger engine with horizontal valves, the exhaust outlet was at the top of the cylinder head facing upwards, and separate short exhaust pipes from each cylinder passed through the roof of the Tank to manifolds outside the hull. This was a great improvement from the point of view of comfort of the crew and obviated the necessity for air-jacketing the exhaust system.

As to the combustion chamber, this was of the bath-tub form, with the sparking-plugs in the centre of the roof of the chamber; thus the flame travel was very short and turbulence was increased. With this form of combustion chamber I felt justified in employing the same ratio of compression as on the smaller cylinders of the 150 h.p. engine, namely, 4.3 to 1. In the event I could safely have employed a higher ratio of compression, for unlike the 150 h.p. engine we had no trouble whatever with detonation even on the very low grade fuel which, as we now know, had an octane number of only forty-five.

Meanwhile the 150 h.p. engine was proving adequate in power for the new Mark V Tank so there was not the same urgency for the larger engine and we had time to build a single-cylinder unit. This was constructed by Mirrlees and a second similar unit by Gardners. Both these single-cylinder units gave a performance of rather over 40 h.p.

The design of the six-cylinder unit was completed during the summer of 1917 and orders were placed with Messrs. Gardner, Hornsby and Mirrlees for a batch of these engines to be built concurrently with the 150 h.p.

The number of engines for Tanks turned out during the years 1917–18 was far greater than the number of hulls because these engines were in great demand for a large number of other purposes. Several hundred of the 150 h.p. engines were used in France for providing power and light to base workshops, hospitals, camps, etc. These engines, in many cases, were called upon to run for very long spells. This, of course, was a far more arduous duty than service in the Tanks. Others were in demand for the Navy for the propulsion of all kinds of auxiliary craft. Others yet again were used in improvised shunting locomotives. The Navy, however, preferred the larger 225 h.p.

To sum up, I think the success of these engines stemmed largely from the fact that they were by far the most powerful petrol engines, apart from aviation engines, available to the Allies. It stemmed also, and above all, from the generous help and co-operation on the part of everyone concerned.

At some time in the winter of 1917–18, General Ellis, then in command of the Tank Corps in France, invited me to visit him at his headquarters to see and hear at first hand how my new engines were behaving in operational use. I was taken round the Tank repair-shops by the major in charge; this was a vast establishment some fifteen miles behind the fighting front. Power and light were supplied by several of my 150 h.p. engines. Timidly I enquired how my engines were behaving and what troubles they were experiencing. He had evidently been warned of my approaching visit and intended to pull my leg. 'Oh,' he said, 'they behaved quite well until they blew up.' This sounded rather horrifying. I asked him exactly what he meant by 'blowing up'. With a wave of his arm he said, 'Oh, just flying to pieces like a bomb or a shell.' He then beckoned to a sergeant who wheeled in a hand-cart loaded with bits and pieces of what I recognised had once been one of my engines. Of its massive cast-iron bedplate nothing remained but a heap of small broken bits; its crankshaft was tied into a lover's knot with bits of connecting-rod adhering to it. The cylinders, however, were not so badly shattered as the rest of the engine. Never before had I seen so complete a wreck. I asked the major what had happened. He remarked airily, 'Oh, they all go to pieces like that sooner or later.' That was all I could get out of him but I had noticed a broad grin on the face of the sergeant, so I dropped the subject. After that I went all round the repair-shops which were very well equipped. There I saw many more of my engines in various stages of overhaul from which I learned a lot as to their actual behaviour in operational service. Accidents and failures there had been, of course, but there was no evidence of any endemic troubles to cause me alarm. Later that day a young Tank officer knowing that I had been round the repair-shops asked me if I had seen the 'horror'. I told him I had indeed. 'Did they tell you how it happened,' he asked. I said 'no' but that I guessed from the sergeant's grin there had been some funny business afoot. He then told me that it was their practice, before going into action, to place a very large demolition charge under the engine bedplate with which to wreck the machine in case of abandonment, and that one of these charges had been fired by mistake!

CHAPTER 12

Formation of my Company

While the first two years of the war had been for me a period of misery and frustration, of idleness and disappointment, the year 1917 was crowded with events each and all of which had a profound influence on my future career. The year 1917 was also the transition of my status from that of an enthusiastic amateur with a bee in his bonnet about the evils of detonation and the virtues of turbulence, to that of a responsible person whose experience and judgement should be listened to with respect. It brought me also other strokes of good fortune in the shape of a reunion with both my colleagues, Hetherington had Thornycroft. During the first two years of the war Hetherington had been stationed at Farnborough and allotted the task of testing, and reporting on enemy aero-engines. This he had done with his accustomed thoroughness, and had returned to his home in London to prepare a statistical analysis of his findings, as to which he kept me informed. At the same time Thornycroft had transferred from Cave-Browne-Cave's R.N.A.S. Squadron to Squadron 20, and taken charge of the testing of the latest developments in Tanks at the newly formed testing station at Dollis Hill in the London area. He, too, kept me informed of all his observations. He was a most capable and observant experimenter and a first-rate critic and when my new Tank engines came under his charge, I could always rely upon the accuracy of his observations, and had the benefit of his criticisms as to their behaviour and their shortcomings. Thus I had every advantage in my favour.

In the winter of 1916–17 my grandfather had died at the age of eighty-nine. His death brought home to me that in fairness to Palmer and Tritton I must make up my mind what I really wanted to do when the war was over. The offer of a partnership in the firm was open to me but by that time I had fallen so deeply in love with the internal combustion engine that I could not bear the thought of parting from it. I could not see, however, what part research and development work on it could play in the firm's affairs. Again, although I was, on paper, a fully qualified civil engineer, in reality I knew little or nothing about it. I could not see, therefore, how, as a partner, I could take a useful part in the principal affairs of the firm.

On the other side of the picture my appetite had been whetted by the success of the research I had carried out before the war in my workshop at Walton, which was really a continuation of that which Hopkinson had initiated. I had confided my dream of establishing a laboratory of my own in which to carry out research, design and development, to a number of leading engineers in industry, such as Windeler, Chorlton, Bryant and others, and had been assured that such an establishment would have their support, and that they thought I would have little difficulty in finding the capital necessary. Also, my head had been turned by the praise I had received for my design of an engine for Tanks, even though at that date it was still only on paper. I talked all this out with Palmer and Tritton who were kind enough to say that they would be sincerely sorry to let me go, but that they quite appreciated and sympathised with my point of view. I had, of course, at various times talked about this, my ambition, to several of my more intimate friends. Hetherington and Thornycroft both said that they would like to join me in such a venture, as also did Halford. Since my first visit to Vauxhall I had met Kidner on a number of occasions and had told him of my ambition. He had said that such an establishment as I envisaged was sorely needed, and that I could count on his firm to retain us as consultants. Though several others had promised their support, Kidner was the first to do so in definite terms. It was a most encouraging start.

As to the capital required, Dugald Clerk had told me that his friend, Campbell Swinton, made a hobby, as he expressed it, of financing promising young inventors. Campbell Swinton, he said, was a very distinguished consulting electrical engineer who had been elected to a Fellowship of the Royal Society on account of his pioneer work on wireless telegraphy; also he had a wide circle of friends in the world of finance who relied on his sound judgement. He told me that Campbell Swinton and his friends had provided Sir Charles Parsons with the funds to build his experimental boat *Turbinia* which had put up such a sensational performance at the Royal Naval Review some twenty years earlier, and thus established the steam turbine as the marine steam engine of the future. He had also financed his friend Marconi on some of his developments. I said I would feel very shy of asking so great a man for the money to build the laboratory of my dreams, but he replied that if I liked he would sound Campbell Swinton about such a proposition. I recalled that some seven or eight years earlier Dugald Clerk had introduced me to Campbell Swinton when he took me to lunch at the Athenaeum when we had talked about the lack of independent research establishments in this country, and that Campbell Swinton had expressed the view that it would be much more difficult to find the right man to run such an establishment than to find the necessary finance.

Dugald Clerk, it seemed, spoke to good effect, for Campbell Swinton sent for me to come to his office and asked me to tell him what I had in mind, and how I proposed to set about the project if the necessary capital were available. I told him that it was my ambition to establish a laboratory in which to carry out long-range research on the internal combustion engine and its fuels, and to design, construct and develop prototype engines on behalf of industry. After I had explained my plans in detail he said that I seemed to have got it all thought out and that he would like to talk it over with some of his friends who might be interested. This meeting must have been in February 1917. The design of my Tank engines had been completed and I was waiting on tenterhooks for the first engine being rushed through by Brotherhoods to be ready for test, always an anxious time with any new design. At that date, also, we were just starting work on my supercharged aero-engine which hitherto had been bandied from pillar to post, and which, but for Halford's tenacity, would probably have been abandoned altogether. I felt, therefore, that if either of these projects proved a success I should have a much stronger case for my request for capital for my laboratory.

Campbell Swinton was a bachelor, at this time aged about sixty, who lived alone in a magnificent house in Belgrave Square, in the basement of which he had his own electrical laboratory devoted

mainly to his researches into wireless communication of all kinds
on which subject he was generally regarded as the leading authority in this country. He was, I believe, a wealthy man, derived
partly from his fees as a consultant, and partly from his acumen
as a financier; he had a wide circle of friends both among his
Fellows in the Royal Society and in the world of finance. In peace-
time it had been his pleasure to give select little dinner-parties to
his more intimate friends, such as Sir Charles Parsons, Dugald
Clerk, Ferranti, Marconi and others in the world of science, or of
finance, such as Goodenough, then Chairman of Barclays Bank.
Though during the war food rationing greatly restricted the menu,
he still clung to the Victorian splendour of his table, with its superb
silver and glassware, and served by his immaculate butler. During
1917 and the years that followed, I was invited to several of these
dinner-parties, and it was at one of them, over the nuts and wine,
that my present firm came into being.

During the winter of 1916 and the spring of 1917 the war had
reached its deadliest phase; to all its other horrors had been added
the use of poison gas, while the unceasing heavy shelling had
broken down most of the locks of the canals and irrigation
ditches and flooded wide areas of our fighting front into a sea of
mud. Through this our men had to struggle as best they could,
and many seriously wounded were unavoidably left to drown, for
rescue work was almost impossible. To the long daily lists of dead
were added other columns under the heading of 'Missing', and
one shuddered to think what that term might imply.

Our ally, Russia, was tottering on the brink of revolution and
there were rumours of mutinies among the French forces on our
right flank. At sea, the ever-increasing fleet of German submarines was taking an appalling toll of our supply ships, despite the
introduction of the convoy system in which they were escorted by
destroyers and other naval vessels. As a consequence, food rationing, especially for civilians, was becoming severe and, for the
first time in my life, I learned what it was like to be really hungry.

At long last the Admiralty had changed its allegiance from
airships and balloons to long-range flying-boats similar to those
envisaged by Commander Briggs some two years earlier and which
did, in fact, prove to be our most effective weapon against enemy

submarines during the last year of the war. I was fortunate, however, during that gloomy period in that, at long last, I had stumbled upon what really seemed a worth-while job and one which occupied the whole of my thoughts and energies for most of that year. As to my dream laboratory the meetings I had had with Campbell Swinton at his London Office had left me with the definite impression that he was interested in and favourable to my project, but I did not think that anything could be done about it until after the end of the war, and there was no end in sight.

I have said earlier that my new 150 h.p. engine was rather prone to detonate on the very inferior petrol allocated to the Tanks, and some time in the early spring of 1917, I was asked to attend a high-level committee which allocated the supply of fuel to the various Services. Most of the members of this committee were high-ranking naval and military officers, but the chairman, a huge and formidable looking fellow, was a civilian. All present were complete strangers to me and in vain I pleaded for a better fuel for the Tanks. That which was allocated to us was then known as 'U.S. Navy Gasoline'. Expressed in modern terms, it had an octane number of only 45, and its volatility seemed only a little better than that of kerosene. I was given to understand that the best quality petrol was earmarked for aviation; the next best for high-speed staff cars, and the lowest grade for tractors and heavy vehicles, and that the Tanks, which only waddled along at walking pace, would have to be content with the dregs of the barrels. Next, I pleaded to be allowed benzole of which, I understood, enough was available for our use. At this suggestion they held up their hands in horror, and said it was quite out of the question since benzole, because of its high specific gravity, was ruled out as an acceptable fuel for petrol engines. They went on to ask why I wanted benzole. I explained that because it was not inclined to detonate, its use would allow me to raise my compression ratio and thereby score a considerable gain both in power and economy and thus radius of action, but they would have none of it.

I came away from that meeting utterly defeated, but on leaving the chairman took me aside and said 'what's all this stuff about benzole and detonation; I would like to hear more about it'.

From the secretary I learned that the chairman was Sir Robert Waley Cohen, then Managing Director of the Shell Company.

A few days later I got a letter from Sir Robert inviting me to dine with him at his home at Highgate. There he examined me about fuels and I told him about my work with Hopkinson, and about the more recent experiments I had carried out in my own workshop before the outbreak of war, which had convinced me that it was the incidence of detonation which limited the compression ratio of the petrol engine, and therefore its power and efficiency. I said that I was satisfied that it was a phenomenon quite distinct from pre-ignition, and that it was dependent upon the composition of the fuel. I told him also about my single-cylinder supercharging engine with variable supercharge which could be used to serve as an approximate knock-rating test unit, and of my experiments on small samples of paraffins, napthenes and aromatics prepared for me my Dr. Ormandy. He said he would like to send me samples of petrol from the company various oil fields for me to evaluate in terms of their tendency to detonate. He said, also, that he would like to send Mr. Kewley, his Chief Chemist, to Walton to witness the tests.

A few days later there arrived from Shell Haven about half-a-dozen five-gallon drums of petrol, each marked with an indentifying number. I tested samples from each of these drums and was disappointed to find that, in most cases, I could detect no significant difference in their tendency to detonate. One sample alone, however, stood out far and away better than any of the others. With this fuel I could utilise far more of my supercharge air than with any of the others, and thus gain at least 20% more power. Kewley and I made several more check tests, both in my supercharging engine and in my Dolphin at its high compression ratio. Neither Kewley nor I knew anything about the origin of this particular petrol, whose behaviour resembled much more nearly that of benzole, and we wondered whether some of the latter had accidentally found its way into this particular drum. We therefore had a fresh supply sent from Shell Haven which we tested, with the same result. In the meantime, Kewley had examined a sample in his laboratory at Fulham, and had found that its specific gravity was much greater than that of the other samples. His analysis revealed that this particular fuel contained an exceptionally large proportion of aromatics, which accounted both for its good performance and its high specific gravity.

We reported our findings to Sir Robert who told us that this particular sample was from a large oilfield in Borneo, and that tens of thousands of tons were being burned to waste in the Borneo jungle because, like benzole, its specific gravity was too high to meet any specification. This was a shocking revelation.

Sir Robert, I understand, cabled at once to Borneo to stop the wastage and arranged for it to be transported to England to be blended with petrols from other sources, and thus improve the quality of their product as a whole. Thus began my long and happy association with the Shell Company.

It seemed that while I was so deeply immersed in the development of my new Tank engines, Campbell Swinton had been in touch with his friend Tennyson D'Eyncourt, from whom he had learned about the progress of my engine. D'Eyncourt, I gather, had reported very favourably and had told Swinton how much he had been impressed by the warm reception accorded me by the would-be manufacturers at our first meeting under his chairmanship in the previous October, and by the smoothness and efficiency with which the whole project had so far been carried out. Swinton, I gather, had told D'Eyncourt of my request for capital wherewith to establish a laboratory for research and development and D'Eyncourt had expressed the opinion that it would be a good investment. On the strength of this and other enquiries he had made, he had invited two of his friends, Goodenough, chairman of Barclays Bank, and De la Rue, chairman of the firm of that name, to share with him in putting up the capital needed for such an enterprise. He went on to say that both Goodenough and De la Rue would like to see me and to hear what I had to say about it. To this end he invited us three, and also D'Eyncourt, to dine with him at his home in Belgrave Square.

The dinner-party that evening was not so embarrassing as I had feared. Campbell Swinton was a good host and did his best to put me at ease. At his request I repeated what I had told him at our earlier meeting both of my experiences since my Cambridge days, and my suggestion as to the sort of laboratory I had in mind, where it might be located and from whence I hoped the revenue might be derived. I left it to D'Eyncourt to tell the story of the engines for Tanks. D' Eyncourt said that his committee had entrusted me both to design these and to find a group of manufacturers to produce them. He went on to say that in a very short time a prototype engine had been completed which had already undergone some weeks of drastic testing, developing rather more power than I had led him to expect, and that a first batch of production engines was almost complete. The future of the Tank in this war depended largely upon the success of my engines, for they had no other string to their bow. He ended by saying that what had led him to entrust so great a responsibility to me was the warmth of the reception of my proposed design, and the evidence of goodwill displayed by all the would-be manufacturers when the project was first put to them some six months earlier.

De la Rue took the line that my laboratory proposition would be unique and highly speculative in that its success or failure would depend on our ability to acquire and maintain the confidence and support of industry. In this connection he had been much encouraged by what D'Eyncourt had told him, but none the less he would like to postpone any decision as to whether he personally would participate until the success or otherwise of my new Tank engines had declared itself, for he gathered that both D'Eyncourt and I were still keeping our fingers crossed.

Goodenough agreed that the whole project was, of course, highly speculative. As he saw it, the proposed company would consist of two groups, the capitalists and the technical group, the latter consisting of myself and my friends Hetherington, Thornycroft and Halford, all young men in the early thirties, with careers still ahead. If the project succeeded it might well prove very profitable to both groups. If it failed, the capitalists would lose the money they had invested, which would be bad but not calamitous, but for the technical group failure would be a major disaster, for they would be left out in the cold at the threshold of their careers, and with the stigma of failure to live down. He urged me, therefore, once again to ponder deeply on whether I, a married man with a family, and with the promise of a partnership in a long-established and very prosperous firm of civil engineers, would be justified in taking such a risk. He went on to say that he agreed with De la Rue that we should wait to see the behaviour of the first batch of Tank engines due for completion in a few weeks' time.

We then discussed how much capital would be required to build and equip a laboratory such as I had planned and generally to get the whole scheme under way. It was estimated that we should need a capital of between £20,000 and £30,000. Both De la Rue and Goodenough agreed that they now had a pretty clear picture of what I had in mind, and that they would consult other friends in the world of finance who might be willing to participate, and that we would meet again in about six or seven weeks, by which time the position *vis-à-vis* my Tank engines would be clarified. In the meantime I would keep Campbell Swindon informed of any new developments bearing on the situation. Those few weeks were fraught with several vitally important developments. The first batch of production Tank engines had been completed by the group of Manchester firms, and thanks to the experience and data derived from our tests on the pilot engine at Brotherhoods, they had passed their test with flying colours, both on the test-bed and as replacement engines in several Mk. IV Tanks; thus their success seemed assured. With the prospect of my company now rather nearer, Sir Robert Waley Cohen, who had previously promised support, now told Campbell Swinton and myself that the Shell Company were prepared to spend up to £10,000 per annum for a period of three years on a fundamental research into the problem of matching fuels to the needs of engines, and that he hoped that this would give a flying start to our enterprise. This was indeed a generous offer, on the strength of which Campbell Swinton organised another dinner-party, this time to discuss not whether but when and how to get going with our enterprise.

At this second dinner-party the guests included Goodenough, De la Rue, Staples, Beale and Brooman-White, the three latter being strangers to me but friends of Campbell Swinton. I cannot remember whether Lord Combermere was present that evening – he had been on active service at the front in France and had been badly gassed and invalided home. All these, including Lord Combermere, had expressed their willingness both to take shares in the Company and to be members of its board. In addition, De la

Rue had brought with him a Mr. Campbell Farrer, senior partner in the firm of Farrer, Porter & Co., a firm of solicitors who specialised in company law and the formation of new companies and who had acted for De la Rue on previous occasions. I recalled that I had met Farrer before the war when he was acting on behalf of a small syndicate whose intention was to market the Dolphin-engined Vox car, made by Messrs. Lloyd & Plaister, but the outbreak of war had put an end to all that.

I had assumed all along that we should have to wait until after the end of the war and the release from service of my technical collaborators before we could take any active steps about the building of my laboratory. To my great surprise all present agreed with Goodenough that we should strike while the iron was hot and get going as soon as possible. Goodenough argued that at best it would take two years to design and build our laboratory, and that sheer exhaustion must bring the war to a close before that. He predicted that the end of the war would be followed by a scramble to convert swords into ploughshares, and a period of drastic cuts in expenditure of all kinds. Under such circumstances it would be difficult, if not impossible, to get anything done, whereas if we got cracking at once, we could rely on the Ministry of Munitions to grant us the necessary permits and take advantage of the prevailing atmosphere of goodwill and offers of support promised us by a number of industrial firms such as that just received from the Shell Company. To this also all present agreed.

We next discussed the name of the firm to be. Several members wanted it called by my name, but both Campbell Swinton and Goodenough raised the objection that if it came to grief from any cause, its failure would inevitably be attributed to my incompetence. They pointed out that the name could later be changed to that of 'Ricardo & Co.', after its success had been established. (In point of fact it was so changed a few years later.) In the meantime they preferred an impersonal title such as 'Engine Patents Ltd.' This, too, was agreed.

As to finance, Campbell Swinton proposed that the capitalist group should put up £20,000 in the form of participating preference shares of £1 each, and that I should be allotted a like number of founder shares of a nominal value of one shilling each but ranking equally for dividend with the preference shares, after the preference dividend had been paid. This allocation as between an inventor and his financial backers had, he said, been found to operate very fairly in the case of other concerns in which he had been interested, and in this both Goodenough and De la Rue concurred. To me it seemed a very generous arrangement. I enquired whether there would be any objection to my handing over some founder shares to Hetherington to whom I owed so much. They replied that there would be no objection provided that I myself retained a substantial majority of these shares.

Several members expressed their doubts as to whether the amount of capital would be adequate, but Campbell Swinton pointed out that until the laboratory was completed, in say two years' time, very little expense would be incurred beyond that on bricks and mortar. Of the would-be technical staff all but myself were serving officers, whose salaries were being taken care of by the Services, while as to my salary, so long as I was on loan to the Ministry of Munitions, I was their responsibility. He considered, therefore, that the proposed capital would be adequate to meet the needs of the foreseeable future.

As to the site for the laboratory, Campbell Swinton said that he himself and an expert had had a look at the site at Shoreham-by-Sea I had proposed and thought it quite suitable. Incidentally this was the site that my cousin Ralph had coveted ten years before whereon to build a huge factory to turn out Dolphin cars by the thousand. All was agreed and final instructions given to Farrer to go ahead with the legal formalities entailed in forming the new Company and to initiate the purchase of the site, while I was to invite my father to prepare designs for the new laboratory. Thus our new company was started on its way far sooner than I had dared to hope. Farrer proved to be a quick worker for all the legal formalities, including the purchase of the site, were soon completed and the company was registered under the title of 'Engine Patents Ltd.', with Campbell Swinton as its first chairman in July 1917.

I had, of course, kept Tritton and Palmer informed of the steps being taken towards the formation of my Company. They

225 h.p. Tank Engines being built at Messrs. Gardner's works

Workshop at Walton-on-Thames, c. 1914

Workshop at Penstone, c. 1925

Sir Robert Waley Cohen
By permission of Shell

Sir Henry Royce
By permission of Rolls-Royce

Tizard as a member of the Experimental Flight, Upavon, 1915
By permission of the RAeS

*Mervyn O'Gorman, from a photo taken three years after his
period as Director of the Royal Aircraft Factory, Farnborough*
By permission of RAE Farnborough

both said again that they were very sorry to see me go, and generously offered to keep the partnership open for me for a period of two years after the end of the war in case my new enterprise failed and I wanted to return to them. My salary, which during my years with Rendel, Palmer & Tritton had risen from £150 to £400 per annum and had latterly been defrayed by the Ministry of Munitions, was discussed and Campbell Swinton and Farrer took the matter up with the Ministry's Director of Contracts. He, it seems, told Farrer that he was shocked to learn that my salary was only £400 which was out of all proportion to the responsibility I was undertaking, and the upshot was that it was fixed at £2000 per annum, back-dated as from October 1916. This was indeed a surprise.

By the end of 1917 therefore I could at last feel that I was playing a really useful part in the war, my dream laboratory was well on the way to becoming a reality, and I temporarily became a relatively wealthy man. In other circumstances I should have been blissfully happy, but over all hung the continuing ghastly nightmare of the war.

In the autumn of 1917 I received a letter from Professor Hopkinson saying that he had been appointed Technical Director of the newly formed Air Service, and as he knew that I had kept up my interest in aero-engine development in this country, he asked me to join him as his part-time deputy. This was, of course, a job after my own heart, and a wonderful opportunity for me, since it brought me into contact with the technical staffs of all the principal firms concerned with the development of aero-engines and also with Farnborough, and other testing stations such as Martlesham Heath. To have as my chief my erstwhile professor at Cambridge, a man for whom I had the greatest admiration and affection, was the culminating stroke of good fortune. My only misgiving was lest the various firms and establishments might resent my intrusion, but far from this being the case I was received everywhere with the utmost cordiality and kindliness. Without the slightest reticence they told me of all they were doing at the moment, and proposing to do as a next step, of the teething troubles they were encountering and the steps they were taking to overcome them. They had no objection to my passing on such information to the other firms I visited, for at this time rival interests were submerged in a common purpose.

I paid frequent visits to the Royal Aircraft Establishment at Farnborough where early in the war O'Gorman had established a new department for engine testing and for research into the various problems of engine design and development. I was tremendously impressed by the really excellent work this department was carrying out under the able direction of Norman. In my memory Norman remains as an outstanding personality who combined inventive genius with sound practical judgement. He was never didactic but had a simplicity and charm of manner which won him the affection and esteem of all his staff. His sad death shortly after the war was a grievous loss to the country. Among his able and well-chosen staff was Professor Gibson, who was carrying out a comprehensive research into the problem of air-cooled cylinder design. He was, I believe, the first to develop a practical design of aluminium cylinder head, and to solve the problem of how to attach it to the cylinder barrel without interfering with the flow of coolant air. His work made possible the use of air cooling for engines of really large power. When I visited the U.S.A. in the spring of 1922 I was interested to find both the firms of Pratt and Whitney and Wrights relying on Gibson's findings, and employing his method of cylinder head attachment for the large radial air-cooled engines they were both developing. Like Norman, Gibson was a man of great personal charm and simplicity and we became great friends. Other members of this Department were Green and Ellor. Green had been primarily responsible for the design and development of the R.A.F. engine, virtually the only British engine available at the outbreak of war.

Looking back over the years I am more than ever impressed by the excellent work on engine research and development work carried out at Farnborough at that time, and I feel that it has never received the public recognition it deserved.

I found my visits to manufacturing firms and Government establishments extraordinarily interesting. The personalities who stand out in my memory are Rowledge of Napiers, Hives of Rolls-Royce (later Lord Hives), Fedden of Bristol and Tizard of Martlesham Heath. The more I saw of Rowledge the more I liked and respected him for his sound judgement, his ingenuity and intellectual integrity. He was by nature shy and modest about his own achievements. These I learned about from other members of his staff, all of whom held him in great respect, as did I when I came to know him well, and we remained great friends until the day of his death.

At Derby I used to meet Hives and members of his staff. Hives at that time was in charge of all engine testing at Rolls-Royce. He was always very welcoming to me and taught me a lot about the mechanical problems met with in their test-shop. He struck me as an essentially practical engineer, a tower of strength, of good common sense, and a loyal disciple of Royce whom he regarded as a real genius. As to Fedden, he was always very nice to me both then and in later years when we were engaged on the development of the sleeve-valve engine.

With Tizard who, at that time, was in charge of experimental flight testing at Martlesham Heath, I found that I had many interests in common, and almost at once we developed a very warm friendship that endured to the end of his life. Tizard, like myself, believed that there was a very real need for a thorough investigation into the properties of the available fuels and their influence on the behaviour of the engine. He was, I think, the most brilliant of my contemporaries, a delightful companion, with a pretty sense of humour.

To my regret I never had an opportunity of meeting Royce during that period, but in the years that followed I came to know him well. He was by far the most brilliant mechanical designer I have ever met; he was also the finest example of intuitive genius, an artist to his finger-tips and a perfectionist. He had the reputation of being somewhat of a dragon, who breathed fire on anyone at Derby who dared to interfere or modify any part of his design without his approval. All his staff, I gather, both feared and worshipped him.

During the 1920's I used to meet him more frequently both at his home at West Wittering, where he lived all summer, and at his villa in the South of France in winter. He was always charming to me and to my children who loved him. His gracious manner captured us all. He discussed his engine designs with me and listened to any suggestions I might have to make on technical matters almost as though I was a fellow artist, which I found very flattering.

My work as unofficial deputy to Hopkinson occupied only part of my time, but it brought me into contact not only with the personnel of the many firms engaged on aero-engine production, but also with the leading technical authorities in the Ministry. Hopkinson himself was kept very busy with administrative work and generally left it to me to deputise for him on the Aeronautical Research and other committees such as that dealing with fuels for aero-engines. Thus, by the end of the war, I was as well informed as any other single individual on the state of the art, not only in this country but also, thanks to Hetherington's coaching, in Germany as well. Thus, I was able to keep Hopkinson posted and, at the same time, to spread his gospel, that the crying need was for greater and ever greater engine power and lower specific fuel consumption. Preaching this gospel, however, brought to me the realisation that my long-cherished R.H.A. engine, being built at last by Armstrongs, would be out-classed by the time it was completed.

Like O'Gorman before him, Hopkinson had gathered round him a group of young physicists, all of whom were contemporaries of mine, among them Keith Lucas, f.r.s.i., Pye (later Sir David Pye, f.r.s.), Darwin (later Sir Charles Darwin, f.r.s.), Lindemann (later Lord Cherwell, f.r.s.), Tizard (later Sir Henry Tizard, f.r.s.), Taylor (later Sir Geoffrey Taylor, f.r.s.), Southwell (later Sir Richard Southwell, f.r.s.) and Farren (later Sir William Farren, f.r.s.). These young men, popularly known as Hopkinson's 'gang' he distributed to the various testing stations, such as Farnborough, Martlesham Heath and Orfordness. He must have picked well for all of them, except Lucas who was soon killed in an aeroplane accident at Farnborough, achieved high distinction in later life. He picked also, as his personal assistant, Aubrey Evans, who held the war-time title of 'Major' Evans, which, for some unexplained reason, clung to him for the rest of his life. Evans was a year younger than myself. He had worked with Sir Dugald Clerk in his researches on the internal combustion engine during the same years I had worked with Hopkinson at Cambridge, and later in Sir Dugald Clerk's Patent Office, wherein he had accumulated an encyclopaedic knowledge of all that had been done in the past. Evans and I shared the same admiration for Hopkinson. We soon became friends, and after Hopkinson's tragic death in the autumn of 1918, he joined our firm of Engine Patents Ltd. and remained with us until his death in 1960.

CHAPTER 13

A New Direction

During the four years of the First World War my time and thoughts were focussed on the design and development of petrol engines for both Tanks and aircraft propulsion and I was only vaguely conscious of what was going on in other fields of mechanical engineering. My impression is that since all eyes were turned to the development of engines for military use during that period, little attention was paid to industrial engines, or even to the engines for road vehicles in which latter field the accent was on production rather than on improvement. Prior to the war, the design of engines for road vehicles had, with a few exceptions such as Lanchester, Pomeroy and Rowledge, been in the hands of cycle-makers, superb mechanics, well versed in the art of light mechanical design but, for the most part, abysmally ignorant of thermo-dynamics, or of the many other factors upon which the performance of their engines depended; thus it was anybody's guess what this might prove to be. If they were lucky they would be regarded as wizards, possessed of some magic and mysterious secret known only to themselves.

In those early days scientists and research workers were apt to be regarded as freaks, wrapped up in a world of their own, far remote from the realities of life and lacking in practical experience. To all suggestions for improvement, the invariable reply would be, 'Oh, that may be all very well in theory, but it won't work in practice,' and the subject would be closed. The war, however, brought about a complete change of attitude as the result of an intrusion of scientists into industry who quickly debunked the prevailing witchcraft and set to work to rationalise the petrol engine. It was, I think, in the aero-engine field that their influence was most apparent. At the outbreak of war we in England had only one British-designed engine fit for military service, the 140 h.p. R.A.F. engine designed and developed by O'Gorman's young engineers at Farnborough.

198

For the rest we had to make do with smaller French engines. Pitted against us were German engines mostly of 200 h.p. and over. Although the specific weight of the German engines was somewhat greater than ours this was more than compensated for by their lower fuel consumption, while their greater power enabled their machines to achieve both a higher speed and to carry a much more formidable armament than ours; thus we were, at first, hopelessly outclassed. By the third year of the war we were producing aero-engines such as the Rolls-Royce 12-cylinder 'Eagle' of 360 h.p. which was both lighter in specific weight and of considerably higher power than any of the German engines, and with which we were gaining supremacy in the air. At the time of the Armistice in 1918, we had in an advanced state of development such engines as the Rolls-Royce 'Condor' of 500 h.p., the Napier 'Lion' of about the same power, and the Bristol air-cooled radial engine, rated, I believe, at about 400 h.p. None of the successful British engines developed during the war comprised any startling novel features. They represented, however, step-by-step development along orthodox lines, making full use of the research carried out at the makers' works, at the universities, and at the excellent engine research establishment at Farnborough. Thus, starting almost from scratch, we had caught up with and passed the Germans.

As the war proceeded so the need for supercharging became increasingly insistent. No one as yet had succeeded in developing any form of positive displacement blower suitable for this purpose. In France, Professor Rateau had developed an exhaust turbine, direct-coupled to a high-speed centrifugal blower and this was successful up to a point, but at the low compression ratios to which we were condemned the exhaust was too hot for the materials then available for the turbine. Moreover, the festoons of red-hot exhaust pipe from each cylinder to the turbine was considered too alarming for the pilot. Rateau was in advance of his day, but his achievement set others to work to devise a very compact mechanically-driven centrifugal blower. This, although no easy problem, was eventually achieved, though not in time to take any active part in the war.

I have referred only to the engines for they are all I really knew about, but equally rapid advances took place in the design of airframes and instrumentation in which field scientists played an even larger part. Of Hopkinson's 'gang' of young physicists, most of the members I have named were authorities on aero-dynamics rather than on engines.

After the war one of the activities of my newly formed company was the improvement of engines for automobiles, but to explain this I must return briefly to earlier days.

During pre-war years the most popular form of engine for motor vehicles had been the four or six-cylinder side-valve type. This had many advantages both from a manufacturing and maintenance point of view, and had for the user much in its favour. It had the reputation of being smooth running and silent; it was cheap to make and compact, free from oil leaks and, in general, easier to maintain and more reliable than engines with overhead valves. However, contemporary overhead-valve engines gave a markedly better performance both in fuel consumption and power output. With the ever-increasing demand for more and more power, the popularity of the side-valve engine was beginning to wane. I was not convinced that the difference in performance between the two types need be so great, and I was determined to investigate this but the outbreak of war prevented my doing so for the next five years.

While our new laboratory at Shoreham was under construction my workshop at Walton had been extended to more than double its original size; I now employed, in addition to my assistant, Doughty, three more first-rate mechanics, Pike, Brown and McKechnie, and immediately after the Armistice Major Evans, the first of my collaborators to be demobilised, joined me. Among other equipment we acquired a small four-cylinder side-valve engine to which we fitted Hopkinson's indicator. From tests on this engine it at once became apparent that the rate of pressure rise after ignition was very low, due presumably to lack of turbulence. It was also prone to detonate even though its compression ratio was barely 4 to 1.

In my Dolphin engine with its separate combustion chamber communicating with the cylinder through a restricted neck, turbulence was intensified during the compression stroke; by varying the diameter of this neck we could increase turbulence within the combustion chamber and thus speed up the rate of pressure rise, as shown both by the indicator diagram and confirmed by the reduction in spark advance required. These observations suggested that it might be possible to supplement the initial induction turbulence during the compression stroke in the side-valve engine much as in the Dolphin. Originally the combustion chamber of the side-valve engine took the form of a flat slab of uniform depth extending over the whole of the cylinder bore and the valves, thus involving a very long flame travel, a condition which we knew would be conducive to detonation. We hoped that by concentrating the whole of the clearance volume over the valves communicating with the cylinder through a somewhat restricted passage, turbulence would be increased and, as to detonation, which we believed to be caused by the spontaneous combustion of some end gas entrapped between the piston and cylinder head, we hoped that by bringing these two flat and relatively cooled surfaces into the closest possible proximity, the thin layer of unburnt gas would be so chilled as to escape spontaneous combustion from the advancing flame front. In this connection I recalled that some five or six years earlier when investigating the cause and mechanism of detonation I had fitted to the combustion chamber of my Dolphin engine, in a position most remote from the sparking-plug, a blind-ended steel thimble of about .6 inch bore and about $1\frac{1}{4}$ inches in length to serve the dual purpose of increasing the length of flame travel nearly 50 per cent and act as a sort of hideout for the end gas. To ensure against pre-ignition, the thimble was water-cooled. Thus fitted the engine suffered severe detonation despite the fact that the additional capacity of the thimble lowered slightly my compression ratio, and this supported my belief that detonation was initiated by the spontaneous combustion of a small part of the working fluid ahead of the flame front. As a next step I had inserted through the blind end of the thimble a copper rod of about .4 inch diameter for the full length of the thimble, leaving an annular space of about $\frac{1}{10}$ inch width : the outer end of this rod was water-cooled. Thus fitted the entrapped gas had ample opportunity to get rid of the heat thrust into it by the advancing flame front, and there was certainly no more detonation than with the thimble removed. On the evidence of this earlier observation we felt reasonably hopeful of quenching any attempt at spontaneous combustion of the end gas in a region remote from the sparking-plug by the method we proposed.

We set out to design and make a new cylinder head for the little side-valve engine. To save pattern-making and time we made this in the form of an open top tray covered by a sheet-metal plate. In our first version the combustion chamber was roughly triangular in plan with its apex overlapping the cylinder bore to form a restriction through which the working fluid would have to pass both on its way into the cylinder and back into the combustion space. We realised, of course, that by increasing the turbulence in this manner, we should have to pay the price of increased heat loss by convection, and that we should have to find the best compromise between these two conflicting factors. It was obvious also that the restriction or area of overlap must not be so small as to penalise the volumetric efficiency of the engine.

In our first experimental cylinder head the cavity forming our combustion chamber had vertical side walls. The metal sections of the head were left very thick so that they could be machined away either to enlarge the area of overlap, or to increase the compression ratio.

As a first step we made the compression ratio about the same as that of the original combustion chamber and the area of overlap about half that of the inlet port. Our first tests were rather disappointing in that the running of the engine was rough and noisy. We got little or no gain in power or fuel consumption but there was no trace of detonation even with the ignition timing advanced well beyond the optimum. Instead the engine made a dull rattling noise resembling that of a kettle-drum. Our indicator diagrams revealed that the rate of pressure rise after ignition was very steep in the neighbourhood of 60 or 70 pounds per square inch per degree of crank angle, or more than three times that with the original combustion chamber. We noted also that the optimum ignition timing was barely 10 degrees early as compared with well over 30 degrees early with the original head. We concluded that we had greatly overdone the the degree of turbulence and we therefore increased the overlap or throat area by about 50 per cent by milling back the wall around the apex of our triangular combustion chamber. This gave us a marked improvement both in power and fuel economy, and a reduction in the kettle-drum noise. We continued step by step to enlarge the area of the throat until we reached and passed the optimum area which appeared to be about equal to that of the inlet port. Under these conditions the kettle-drum noise almost disappeared, and the rate of pressure rise after ignition was in the region of 35 to 40 pounds per square inch per degree of crank angle.

Our next step was to increase the compression ratio by machining back the lower face of the head until the incidence of detonation set a limit. The normal ratio in this small engine of $2\frac{3}{4}$ inch bore was just below 4 to 1. To our delight we found that we could increase the ratio to very nearly 5 to 1, thereby scoring a further gain of about 10% in power and fuel economy. The nett result of these changes in both turbulence and compression ratio was an overall gain of about 20% in both power output and fuel economy, which brought the performance of this side-valve engine almost up to that of an otherwise similar overhead valve engine.

I have enlarged on these tests carried out in my temporarily extended workshop at Walton-on-Thames immediately after the end of the war because the royalties derived from our so-called turbulent head formed our largest single source of revenue during the next fifteen years, and also because in the patent action in which we became involved in 1933 this original experimental cylinder head was accepted in court as evidence that the optimum area of throat had been arrived at by a step-by-step enlargement as indicated by the tool marks, and not merely fortuitously as the opposition contended.

As a next step we modified our pattern to eliminate the right-angled corners and reduce, as far as possible, the surface to volume ratio. Our modified combustion chamber had a slightly domed roof with a taper lip extending over the cylinder bore. This was an improvement. At the same time it had the effect of reducing slightly the rate of pressure rise and eliminating altogether the kettle-drum noise, and had no overall adverse effect on the power output or fuel consumption.

Evans and I were delighted, for it seemed we had found a very simple way of improving the performance of a conventional side-valve engine by as much as 20%. A few years later we carried out in our new laboratory at Shoreham a more detailed and comprehensive research on the process of combustion in this form of cylinder head. For this purpose we employed a much larger single-cylinder engine into the head of which we fitted a row of glass windows extending from as near as possible to the sparking-plug to the far side of the cylinder bore. Above the cylinder head we fitted a large diameter metal disc with a single narrow radial slot to serve as a stroboscope. This disc was driven from the cam-shaft by spiral gearing in such a manner that its phase relation could be varied while running, and the spread of the flame front from the sparking-plug outwards could be watched and timed, while simultaneously indicator diagrams recorded the rate of pressure rise. In a darkened room the passage of the flame from window to window could be watched. As between the passage of the spark and the appearance of the flame in the first window, there was a delay period of between 5 and 10 crankshaft degrees, after which the passage of the flame across the main body of the combustion chamber was very rapid and accelerating until it reached the thin layer between piston and cylinder head, when it slowed down and did not reach the far side of the cylinder bore until the piston had descended an appreciable distance. If the depth of the layer exceeded about $\frac{1}{8}$ inch, there would be no slowing down of the flame front and audible detonation would occur; as the flame front reached the last window a yellow flash appeared accompanied by the usual high-pitch knock. This was the first time we had been able visually to track the passage of the flame front during its journey across the combustion chamber. Although I do not think it taught us anything new or unexpected, this investigation did provide confirmatory evidence in support of our theory as to the process of combustion and the mechanism of detonation. I found this very convincing, for to me, at any rate, seeing is believing.

In the previous chapter I described the circumstances under which I first came into contact with Sir Robert Waley Cohen early in 1917, and the experiments I had carried out in my workshop at Walton-on-Thames involving the petrol from Borneo. Sir

Robert now implemented his promise to give us a contract to investigate fuel quality, and during the months that followed I had many meetings with him and his Chief Chemist, Kewley, during which we discussed at length how the projected research should be carried out, and what help and equipment we should need. I told him I thought we should need the help and guidance of a first-class physicist and an organic chemist. As to the former I told him that I would like to enlist the help of Henry Tizard whom I knew to be interested in the subject under discussion. Tizard at that time was in technical charge of flight testing at Martlesham Heath, with the wartime title of 'Colonel'; he was a little younger than myself and had had a brilliant career at Oxford and had studied under Nernst in Germany and, like myself, was one of Hopkinson's ardent disciples. I sounded Tizard on the subject and he told me that while it was his intention to return to his job at Oxford University, he thought he would be able to join us on a part-time basis. He went on to say that he would like to bring in David Pye, who was another of Hopkinson's disciples, and at that time was acting as scientific adviser to the Ministry of Munitions on fuels and lubricants. Waley Cohen was much impressed by both Tizard and Pye when he met them, and agreed that we should retain them both on a part-time basis. On the chemical side, Kewley could and did provide us with all the help we needed. He was a great authority on the chemistry of petrol and petroleum products and of refining processes generally. His help proved invaluable, more especially in preparing or procuring pure samples of the individual petroleum components.

Regarding the equipment needed, we all realised that although my little test engine with its variable supercharge had served to identify the Borneo petrol, we should, for the projected research, need an instrument of far greater precision and one which would define the highest useful compression ratio which could be employed with any given fuel. For this we needed a highly efficient single-cylinder engine in which the ratio of compression could be varied while running at full power, and that without any change in temperature or other working condition. It was essential, we felt, that the proposed test engine should have the highest possible volumetric, thermal and mechanical efficiency. I contended that to be really convincing we needed absolute, not merely relative, performance figures.

As soon as our new Company was formed, and our drawing-office in London established and reinforced by the recruitment of four more draughtsmen, we set to work on the design of just such a variable compression engine, which became known as the 'E35', and upon a single-stroke compression machine which, because of its shape, became known as 'The Sphinx'. The cylinder dimensions of both units were the same. Both these machines were made for us by Messrs. Peter Brotherhood and their completion coincided with that of our new laboratory at Shoreham in the early spring of 1919. In the meantime, Tizard and Pye had prepared a monumental analysis of the physical and thermal properties of all available volatile liquid fuels, setting out not only such obvious data as specific gravity, boiling point, calorific value and latent heat of evaporation, but also the total internal energy in terms of foot lb. per standard cubic inch of a fuel/air mixture in the proportion required for complete combustion. In brief, their analysis revealed that when all the relevant factors were taken into account, the potential power output of a spark-ignition engine running on any fuel derived from petroleum or coal would, at any given compression ratio, be the same to within plus or minus 1%. This, at the time a surprising conclusion, was amply confirmed by our subsequent tests on the E35 engine. The really significant difference between fuels lay in their tendency to detonation. Of all the pure substances embodied in petrol or benzole, we found that the paraffin, normal heptane, was the worst, and the aromatic, toluene, the best from the point of view of detonation of any of the hydrocarbon derivatives that Kewley could procure for us.

We realised, of course, that to express the tendency of a fuel to detonate in terms of the highest compression ratio at which it could be used in our E35 engine would not necessarily apply to other engines with larger or smaller cylinders or different forms of combustion chamber. We therefore decided to express it in terms of the proportion of toluene added to heptane needed to match any given sample of petrol. Thus, we expressed the tendency of a fuel to detonate in terms of its toluene value and, for the next few years, the 'Toluene Number' was accepted by the Ministry and the trade generally as the figure of merit of any fuel for petrol engines. Some years later at the request of the Americans it was agreed to substitute iso-octane for toluene, and the 'Octane Number' of a fuel became accepted all the world over as the figure of merit of any brand of petrol.

I do not propose in these memoirs to go into technical details of our three-year research on behalf of the Shell Company. Tizard had asked Waley Cohen to permit publication of all our results after a suitable delay period. To this he agreed and they were published very fully in the technical press and various text-books, including my own, during the early 20's. I will, however, mention some sidelines arising out of that work.

After our discovery in 1917 of the high anti-knock value of the Borneo petrol I had suggested to Kewley that it might be possible to extract a narrow cut from the more volatile distillation range of that fuel to provide a small quantity of super aviation fuel for special purposes, such as long-range flights. To do so would be both costly and wasteful in that it would mean taking the prime cut from the joint and throwing the rest away, but Kewley agreed to see what he could do by fractional distillation. He did succeed in producing a small quantity of light volatile and highly aromatic fuel but of rather too high a specific gravity to comply with the existing specification for an aviation fuel. I tried some samples of this in my engine at Walton and found it very good. He sent samples of it both to Farnborough and to Rolls-Royce and they found that by its use they could safely raise the compression ratio of the Rolls-Royce 'Eagle' engine from 5 to 1 to 6 to 1, and thereby gain about 10% in power output, and about 12 to 15% in fuel economy at the normal cruising speed. There was not, and probably never would have been, enough of this fuel for general use by the Air Force but shortly after the Armistice we had a visit from Alcock and Whitten-Brown, who were planning to attempt crossing the Atlantic in a Vickers-Vimy bomber, powered by two Rolls-Royce 'Eagle' engines. They had heard of this special fuel and wanted to know more about it and what, if any, were the objections to its use. We told them that we knew of none except its scarcity, and demonstrated on my little supercharging engine the improvement both in power and fuel consumption that might reasonably be expected. They said that even a small increase in power or fuel economy might make all the difference between success and disaster for their enterprise, for it was touch and go whether they could take off with enough fuel for the crossing. Kewley said that now that the war was over, and they were in touch again with the company's large laboratory and refinery in Holland, he felt sure that his company could provide enough for both preliminary test flights and for the crossing itself. As all the world knows, these two brave men accomplished the first direct crossing of the Atlantic by aeroplane in June 1919, and naturally the Shell Company made the most of the fact that this great feat had been achieved with the use of their fuel.

Our investigation into the behaviour of fuels of the alcohol group brought into prominence the important part played by the latent heat of evaporation of the liquid fuel. The calorific value of, say, ethyl alcohol is much less than that of petrol, but its latent heat of evaporation is about three times greater. According to Tizard and Pye's calculations the total heat energy of a standard cubic inch of an air/alcohol mixture was very slightly less than that for a straight hydrocarbon fuel. Other things being equal, the power output returnable from an alcohol fuel should be correspondingly less; in fact we found it to be between 5 and 10 % greater, the discrepancy being due to the lower temperature and therefore greater density of the mixture entering the cylinder. In short we were making use of the high latent heat of evaporation of alcohol to supercharge the cylinder by refrigeration to a degree that more than compensated for the lower internal energy per standard cubic inch of mixture. This observation suggested to us that it might be amusing to concoct a special fuel mixture for racing-cars and motor-cycles. I discussed this possibility with Waley Cohen who had no objection, in fact, he, too, thought it would be rather fun.

Ethyl alcohol, unlike its sister methyl, did not suffer from pre-ignition or detonation, even at the highest compression we could reach with our E35 engine, but because of its poor volatility, cold starting with neat ethyl alcohol was virtually impossible. We had therefore to add a small proportion of a much more volatile fuel for the sake not only of startability but also of distribution in a multi-cylinder engine. The choice lay between methyl alcohol and acetone, and for a variety of reasons we chose the latter. Because of the low calorific value of ethyl alcohol we tried adding a substantial proportion of benzole as a thermal makeweight, while to compensate for the much lower latent heat of the latter, we added between 5 and 10 per cent of water. The presence of a small proportion of acetone served to act as a mutual solvent and formed a stable mixture between these otherwise incompatible components: thus we arrived at a fuel which in our E35 engine showed no trace of detonation or pre-ignition at its highest ratio of 8 to 1 or, expressed in modern terms, at an octane number of at least 100, as compared with about fifty in that of commercial petrol, and about sixty in that of the best aviation spirit.

In those days there was a real fear that the world's sources of petroleum might soon be exhausted, and that we might have to fall back upon alcohol as a fuel. This fear had brought about a close liaison between the Shell Company and the Distillers Company. The latter were the chief suppliers of alcohol, and provided both the alcohol and the acetone for our tests, and there was, I believe, a tacit understanding that if and when it became necessary to fall back upon alcohol, the Shell Company, with their large fleet of tankers and world-wide marketing facilities would look after the transport and sale, while the Distillers Company would take care of the production and processing.

During the early 20's the sport of motor-racing on the Brooklands track near Weybridge had become very popular, and huge race-meetings for every category of cars and motor-cycles were held every week during the summer season. Some of the racing cars, such as the famous 'Chitty-Bang-Bang' had been fitted with high-powered ex-aeroplane engines, and others with heavily supercharged racing engines. The speed of such cars, well over 100 m.p.h., was considered too high for safety even on that highly-banked race-track, and the organising committee had banned the use of superchargers, or of engines of unlimited cylinder capacity. There was, however, no restriction on the choice of fuel and most competitors used the current aviation spirit which limited their compression ratio to a little over 5 to 1. Both Shell and the Distillers Company agreed to prepare and market, at a high price, a super racing fuel to our prescription, and to pay us a royalty of 1d. per gallon sold. When, however, we applied for a patent on our mixture, we found that such a patent would be valid only if the proportions of the ingredients were exactly as claimed, whereas it was obvious to us that they could be varied widely. There was a danger that our patent would afford us little protection against any would-be competitor who, by simple chemical analysis, could easily discover its composition. To our original mixture of ethyl alcohol, benzole, acetone and water we had added about 2 per cent of castor oil for what was then termed 'upper cylinder lubrication' for the benefit of some of the engines with inadequately lubricated valve stems, and Tizard and Kewley put their heads together to find some complex organic substance which would both defy analysis and give to the exhaust a peculiar and characteristic smell. So far as I can remember we used finely-powdered bone meal, a small pinch of which, combined with the castor oil, gave the exhaust a distinctive and, I am afraid, a rather repulsive smell.

As applied to an existing engine without any modification other than fitting larger jets to the carburettor, this racing fuel gave, at high engine speeds, an increase in power output of between 5 and 10%, but when the engine was suitably modified to provide for a compression ratio of the order of 8 to 1, as much as a 30 % increase could be obtained. Its use had also the advantage that its high latent heat of evaporation, most of which took place after its entry into the cylinder, both lowered the cycle temperature and, at the same time, provided much needed cooling to the piston and exhaust valve. For use on the road, however, this fuel mixture was not satisfactory, for its poor volatility involved bad distribution at low engine speed with consequent rough running and sluggish acceleration. To combat this it was necessary to employ a very rich mixture which, together with the low calorific value of the fuel, meant that the mileage per gallon was only about half that obtainable with petrol. As the price per gallon was about four times that of petrol, the real use of the fuel was limited to track racing at Brooklands and to hill-climbing competitions.

Halford, who had joined us as soon as he was demobilised, was very keen on motor-cycle racing and took part in many of the competitions at Brooklands. His own machine was a 500 c.c. single-cylinder Triumph, a well-designed and beautifully made engine with side-by-side valves but generally of quite conventional form. It had rather low compression and was sadly lacking in turbulence, hence its performance was not at all exciting, but it made a good showing in handicap races thanks to skilful handling. Halford had, of course, tried out the various mixtures we had concocted at Shoreham, and had satisfied himself that given appropriate carburettor adjustment, our latest concoction gave him a marked increase in speed, and he was anxious to try it out on the race-track. For the last meeting of the season, I think 1921, he entered his Triumph in its usual category, but with his tank filled with our alcohol mixture, as a result of which he scored an easy win in his class. This, the first trial of our fuel at Brooklands, so delighted all of us that we decided to design and construct a new cylinder, cylinder head and piston for his engine, going all out to get the best possible performance. With this end in view we employed a pent-roof cylinder head with four inclined valves operated in pairs by push-rods from the original valve mechanism. For the material of the head we employed a high conductivity bronze alloy recommended by Commander Bryant, which had the virtue that the valves could seat direct on the bronze without the need for inserted valve seats. With this general layout similar to that employed in my R.H.A. aero-engine, we were able to combine the maximum possible valve area, a central position for the sparking-plug and a favourable disposition of cooling fins. We made also a very light aluminium slipper piston with a domed top, giving a compression ratio of about 8 to 1. We completed the revision of Halford's engine in good time for the opening of the Brooklands season, but unfortunately we did not have, at that time, either a suitable high-speed dynamometer, or an adequate supply of cooling air to brake test it at full power. From such tests as we could make, it seemed that the peak output had been in the neighbourhood of 24 h.p. or 25 h.p. at a speed of about 5000 r.p.m.

At the opening of the season Halford entered the machine in the Open Class in competition with twin-cylinder machines of considerably larger capacity, and scored an easy win, and continued to do so at subsequent meetings. It soon became known that he was using a fuel of secret composition prepared and supplied by the Shell Company and known as 'Shell Racing Spirit' and that a rival fuel, not necessarily of the same composition, was being prepared by the Distillers Company and known as 'Discol R'. Speculation was rife and it was great fun to listen to the diverse theories expounded by the various competitors and their views of the relative merits of the supposedly rival fuels. The cans were of different colour but were filled from the same vat. The general consensus of opinion appeared to be that some form of high explosive developed in secret during the war had been dissolved in petrol or alcohol. The immediate effect of Halford's victories was that most of the competitors had their tanks filled with either 'Shell Racing Spirit' or 'Discol R', and instructions were given on how to readjust the carburettors. The introduction of this fuel brought confusion to the committee responsible for handicapping, and after one season's racing its use, like that of pressure supercharging, was banned, but large quantities of the fuel continued to be sold to enthusiastic amateurs for ordinary road use despite its high cost and other drawbacks. Thus, what started as a joke had proved quite a profitable venture for the suppliers while it brought us a great deal of amusement and a substantial sum in royalties. More important, it brought us a contract from the Triumph Company to redesign their engine embodying the same features, including our patent slipper piston, but with a compression ratio low enough for use with ordinary commercial petrol and with a cast-iron in place of a bronze cylinder head. This new engine, known as the 'Triumph-Ricardo', proved very popular during the next decade and brought us in several thousand pounds in royalties and a great deal of kudos.

By the spring of 1922, our three-year research contract on behalf of the Shell Company had been completed, and we all agreed that, for the time being, there was little more we could do beyond carrying out routine tests, or tests on samples of petrol from fresh sources of supply, and that the next step lay with the chemists to devise new refining processes to improve the anti-knock value of their fuel.

The Rolls-Royce Eagle VIII engine which powered the Vickers Vimy on its Atlantic flight in June 1919
By permission of Rolls-Royce

The Bristol Jupiter engine
By permission of Rolls-Royce

CHAPTER 14

Some Episodes from the New Company

After the conclusion of our research on fuels for spark-ignition engines Tizard and Pye left us, though in the years that followed they paid us many visits and Tizard on several occasions joined us for summer cruises on our boat. In the next few years our company was involved in many other projects.

Our work on fuels was followed immediately by another three-year contract for a research on lubricants. I believe that this research was financed jointly by Shell and the Air Ministry, the former being interested primarily in the behaviour of their lubricants as applied to internal combustion engines, and the latter in the properties of bearing materials in relation to load-carrying capacity, compatibility and wear. For the Shell work we had the whole-time assistance of Barton, a very able young physicist on the staff of the Shell Company, and for the Air Ministry work we had the part-time assistance of Major Carter from the Royal Aircraft Establishment at Farnborough, and the whole-hearted co-operation of Dr. Stanton of the National Physical Laboratory.

It had been agreed that the findings of our research should be pooled, but while Shell had no objection to publication, the Air Ministry refused permission, with the result that except for one paper by Thornycroft and Barton, none of our findings was made public. This was a pity I think for had they been published they might have saved a great deal of duplication of effort in the years that followed. For the purpose of this research we designed and made the test equipment both for our own use at Shoreham and for the National Physical Laboratory, and we learned much that was of use both to us and to our clients in the years that followed.

For many years it had been the firm belief that castor oil, by virtue of its high 'coefficient of oiliness' was the only oil suitable for aeroplane engines.

When starting a new line of research such as that on lubricants, I always tried, at the outset, to visualise what is really meant by terms which, in this case, were 'oiliness', 'boundary lubrication', 'full fluid lubrication' and so forth. I well remember that I pictured an oil molecule as an untidy complex affair armed with tentacles which, on coming into contact with a metal surface, would immediately dig themselves into that surface and so anchor the molecule like a limpet to a rock, and I supposed that the term 'coefficient of oiliness' was a measure of the security of that anchorage. Under conditions of full fluid lubrication such as obtain in a normal bearing, the two limpet encrusted mating surfaces would be separated by a liquid film which, thin though it might be, would none the less be many molecules in thickness. Under such circumstances the moving member would, at all times, be floating on a liquid film. The coefficient of friction would then be extremely low and largely dependent upon the viscosity of the fluid. Under conditions of boundary lubrication the liquid film would be absent and the two limpet encrusted surfaces would be bumping over each other back to back; under such circumstances the coefficient of friction would be many times greater and the heat generated thereby might well be sufficient to break down the molecule. The two mating surfaces would then come into metallic contact, when scuffing or actual seizure would occur. Such, in brief, was the picture I had conjured up from such published information as we had studied at that time.

Dr. Stanton's experiments at the N.P.L. had revealed that so long as relative motion, even very slow relative motion, was continuous it was generally possible to maintain full fluid lubrication throughout all the rotating parts of an internal combustion engine, but that if motion ceased, even for a split second, the supporting oil film would be expelled but would reinstate itself almost immediately after motion was re-started. Thus in the case of a reciprocating piston in a stationary cylinder, there would be a momentary reversion to boundary lubrication when the piston came to rest at either end of its stroke, and the same would apply to an oscillating bearing, such as a piston pin, but that in most cases the period would be so brief that no harm would result. I mention this observation of Stanton's because it has an important bearing on what I have to say about cylinder wear and the single sleeve-valve engine.

I will do no more than mention a few highlights resulting from our joint research. First, and perhaps the most important at that time, was the conclusion that the coefficient of oiliness to which so much importance had been attached was, in fact, of little or no significance as applied to its use in an internal combustion engine. This had been the keystone of the argument of the advocates of vegetable rather than mineral oil, in fact almost the only argument. Our engine tests extending over three years proved the superiority of straight mineral oils in such respects as the formation of heavy carbon deposits and piston ring sticking, while as to wear we could find no difference as between a straight mineral, a compounded oil, vegetable oils such as castor oil, or animal oils such as sperm oil. This conclusion was later endorsed both by Farnborough and by the leading manufacturers of high-speed internal combustion engines. That was the end of any serious use of vegetable oils as engine lubricants, although they continued to be purchased by amateurs for many years.

About 1922–23 Fedden of the Bristol Aeroplane Company was in serious trouble with the crankpin bearing of his new seven-cylinder single-crank radial engine and we were asked by him and the Air Ministry to investigate this as part of our research programme. For this purpose we designed and made a big-end testing machine in which we could, as far as possible, simulate the conditions obtaining in the engine, and Fedden supplied us with a dummy connecting-rod weighted to represent the actual conditions. In the engine the crankshaft was a single piece forging; hence the master connecting-rod bearing had to be split, with some of the articulated rods anchored to the master rod and others to its detachable cap. After many weeks of preliminary work we arrived at a satisfactory test technique and found that with the maximum rate of oil flow that could at that time be tolerated the bearing failed from excessive friction heat at speeds above about 1100 to 1200 r.p.m., which agreed reasonably well with Fedden's own findings in the actual engine. We were not sure whether the high friction heat was due to distortion of the split big-end or merely to the excessively high dynamic loadings. To check this Fedden made up a new connecting-rod with unsplit big-ends. This behaved a little better but not well enough to justify the employment of a built-up crankshaft.

I recalled that many years before Sir Charles Parsons had told me how in his early turbines he had employed, with success, a freely floating bush in his journal bearings and on the strength of which I had done the same in my home-made motor-cycle engine at Cambridge. We had by then also employed it again with success in our E35 variable compression engine and so, as a next step, we enlarged the bore of our unsplit connecting-rod and inserted a freely perforated floating bush. The result of this was somewhat spectacular for with the same rate of oil flow we could run safely at a speed of 1600 r.p.m., the limiting speed of our test-rig.

Both Fedden and the Air Ministry boggled at the idea of employing a built-up crankshaft with all that that would imply in the way of extensive re-design of the engine. They therefore urged that we should explore the possibility of using a split floating bush in a split big-end eye. Stanton was not at all hopeful. He feared that the two halves of the floating bush would take up an attitude inimical to the formation of a fluid film, and so indeed it proved, for both his observations and ours revealed a friction loss higher than with no floating bush at all.

For our own use we had developed a form of built-up crankshaft in which we employed a plain hardened-steel parallel crankpin which we made an easy sliding fit in the webs. The outer ends of the webs were slotted to provide the necessary flexibility to allow them to be clamped on the pin by a bolt.

After much discussion with the technical staff of the Air Ministry, Fedden was given a contract to revise the design of his radial engine incorporating a built-up crankshaft to our design; this, of course, took several months during which we carried out many more experiments on our big-end test rig and on the detail technique of crankshaft design.

When completed, the new version of the engine performed very well indeed. Its maximum speed was no longer limited by its big-end bearings but by its valve mechanism. In his next version, the 'Jupiter' engine, Fedden employed a thermally-compensated valve mechanism which allowed of still higher speeds sufficient to justify the use of a propellor reduction gear. As time went on the introduction of effective oil-sealing rings permitted the use of a much greater flow of cooling oil through the crankpin bearing, and therefore yet higher speeds of rotation, so that never again did this critical bearing set a limit to the performance of radial engines.

The success of our built-up crankshaft led to its adoption both in this country and abroad by almost all makers of high-powered radial engines until the advent of the gas turbine some twenty years later brought the reign of the high-powered piston aero-engine to a close.

During the early 20's almost all automobile engines of post-war design were equipped with aluminium pistons of various forms. Frequent complaints were made of rapid cylinder bore wear for which the new piston material was blamed, the accepted theory being that the relatively soft bearing surface of the piston skirt served as a lap to retain road dust or other forms of grit with which to grind away the bore of the cylinders.

Our research on lubricants and bearing materials did not confirm this nor did our examination of worn cylinder bores which, in every case, revealed that the worn area was limited to that passed over by the piston rings. In few if any cases could we find any measurable wear of the bore over the area traversed by the skirt of the piston alone, thus exonerating the use of aluminium for pistons. In every case that we examined the contour of the wear took the form of a deep groove in the cylinder bore at the point where the top piston ring came to rest at top dead centre. The upper side of this groove was sharply defined while the lower side tapered off gradually. There was usually some evidence of a similar wear contour, though not nearly so pronounced, at the bottom end of the ring travel, while between these two local areas the wear was only very slight.

I recalled that Stanton had demonstrated during our joint research that as soon as relative motion between two surfaces ceased, even momentarily, the conditions would become those of boundary lubrication but those of fluid lubrication would be restored almost as soon as relative motion was resumed.

Our research on the sleeve-valve engine had proceeded far enough to reveal that the rate of wear of the sleeve bore was very slight, nor was there any evidence of localised wear at either end of the piston ring travel, but in that case relative motion between piston and sleeve never ceased, for the sleeve was moving while the piston was at rest. Such small evidence of wear as we could detect appeared to be about equally distributed along the whole length of the ring travel.

From these and other stray observations I pictured that in a poppet-valve engine when the piston came to rest at top dead centre on the firing stroke, and with the full gas pressure behind the topmost ring, only a bare mono-molecular layer of lubricating oil would separate the two surfaces. As soon as the piston began to descend this protective layer would be burnt off by the very high temperature flame to which it became exposed, thus leaving a narrow ring of bare metal entirely unprotected from chemical corrosion and so it would remain throughout the whole of the firing stroke and until the piston returned once again on its idle stroke to smear the surface with a fresh layer of oil. Such at least was the tentative theory I had built up. To confirm it visually we wanted to bring a running engine to a sudden standstill immediately after its firing stroke, that is to say in less than one revolution, but we could find no way of achieving this.

At that date we had on our test bed at Shoreham a large single-cylinder, poppet-valve engine on which we were carrying out tests on behalf of the Admiralty. This engine of $8\frac{1}{4}$ inches bore by $9\frac{1}{2}$ inches stroke had a cast-iron cylinder attached to the crankcase by, as events proved, a very inadequate flange. One day when Evans and I were carrying out tests on this engine at full power there was a terrific bang and the whole cylinder suddenly flew up to the roof of our test-shop and then crashed down to the concrete floor almost at our feet, where it split into several pieces. After we had recovered from our surprise we examined the broken bits and found to our delight that a clearly-defined narrow band of rust had already appeared all round the bore of the cylinder and at a position corresponding to that of the top piston ring at top dead centre. Thus, this accident provided just the evidence for which we had been in search to support my theory.

One of my Cambridge friends, Duff, had recently been appointed Director of the Research and Standardisation Committee of the Institution of Automobile Engineers, supported in part by industry, and in part by a Government grant from the D.S.I.R. of which Tizard was secretary. All three of us were members of the United University Club where we frequently lunched together, and as Duff was on the look-out for some worth-while research for the Institution to tackle I suggested an investigation into the problem of cylinder wear and told him of our accidental observation and of my belief that chemical corrosion played a large, if not a major, part in the trouble. Tizard suggested that sulphuric acid might well be the main culprit but would be damaging only so long as the temperature of the surface was below its dew-point whatever that might be. Meanwhile we urged that Duff should collect statistical information as to the operating conditions under which cylinder wear was most severe. From then on Duff and his colleague Dr. C. G. Williams took over research into this subject which they tackled with great ability, and from their engine tests they concluded that most of the cylinder wear in current petrol engines was in fact due to chemical corrosion and that the moral was as far as possible to keep the engine warm so that the inner surface of the cylinder bore should not fall below the dew-point of the various corrosive substances of which sulphur appeared to be the worst offender.

During the early 20's the automobile industry as a whole urged upon its customers the desirability, whenever possible, of running their engines idle for several minutes after a cold start in order to ensure a thorough circulation of lubricating oil. Duff was able to show that nothing could be worse than this custom for initiating cylinder wear. The immediate effect of his findings was the abandonment of this practice and the introduction of thermostats to hasten, as far as possible, the warming-up period of the engine.

Before the end of the First World War the advent of the high flying aeroplane and the incendiary bullet had, between them, spelt the doom of the airship as a military weapon, and in the years that followed fierce controversy raged as to the relative merits of the airship and the aeroplane as a means of long-range civil transport.

My impression is that in the early '20's informed opinion was, on the whole, in favour of the aeroplane although giant Zeppelin rigid airships had made a number of passages across the Atlantic whereas only one aeroplane had achieved this feat. It was unfortunate that the whole controversy developed into a party political issue in which the Labour Party backed the airship and the Conservatives the aeroplane. The Labour Party seemed to have won the battle and plans were made for establishing a regular air service to India by means of a gigantic rigid airship bigger and better than the Zeppelin. The construction of this huge ship was to be a national enterprise. It was, we were told, to be inflated with helium instead of hydrogen and propelled by diesel engines; thus it was claimed it would be immune from any risk of fire. In vain the experts pointed out at the time that there was not enough helium in the world to inflate it, nor had any diesel engines yet been developed light enough and powerful enough to propel it, but their warnings were ignored by the politicians. To my friend Alan Chorlton was set the task of designing and developing the diesel engines, each of about 600 h.p. and considering the state of the art at the time this was indeed a most ambitious task.

At some time in the early 20's a new company, the Airship Guarantee Company, came into being, financed, I believe, largely by Messrs. Vickers, who later took it over entirely. Its objective was to design and construct a rigid airship for the transatlantic service, to be known as the 'R100'. This great airship, designed by Barnes Wallis whom later I came to know very well, was somewhat less ambitious in that it was to be propelled by the already fully-developed Rolls-Royce 'Condor' engines, each of about 500 h.p. Nor was there any suggestion of the use of any gas other than hydrogen for its lift. Thus two rival projects were started, the 'R101' to be built and owned by the State, and the 'R100' entirely by private enterprise. Never was a project so bedevilled both by party politics and by technical antagonisms, with the result that neither ship was completed until about 1930. Although at the time we were engaged on research on aero-engines by the Air Ministry, it was the Airship Guarantee Co. which retained us to act on their behalf.

In 1922–23 we were asked by the Airship Guarantee Co. to look into the question of possible alternative fuels for airship use. To cross the Atlantic involved the carriage of many tons of petrol which left little scope for an economic pay-load and to main-

tain a uniform buoyancy throughout the whole flight it was necessary to jettison a compensating quantity of hydrogen; thus any
reduction in fuel consumption would be doubly valuable.

In these gigantic rigid airships the hull was sub-divided into
a number of separate compartments each containing a fabric
balloon which, when completely inflated, filled almost the whole
of each compartment. It had been suggested that if a gaseous fuel
could be found whose specific gravity was equal to or slightly
greater than air, one or more of the buoyancy balloons could be
converted into gas-holders. As the fuel was consumed and the
balloon contracted the space it had occupied would be replaced
by air and the buoyancy would remain unchanged throughout
the flight. It was essential, however, that the specific gravity of
the gaseous fuel should not be less than the air which replaced
it or the ship would lose buoyancy during the voyage. Thus this
condition ruled out the use of methane or natural gas.

Of the possible gaseous fuels available at that time ethylene
appeared to be almost the only choice and we therefore carried
out tests on this gas in our variable compression engine but found
that, like methyl alcohol and ether, it had a dangerous tendency
to pre-ignition even at very low ratios of compression which put
its use out of court.

During our earlier research on fuels we had been much impressed by the extremely wide range and rapid burning of hydrogen as a fuel. Attempts had long since been made to use the surplus
hydrogen as a fuel by admitting it to the carburettor of the petrol
engine used for our small airships during the war, but it had been
found that the addition of even a very small amount of hydrogen
to a normally carburetted petrol-air mixture provoked violent detonation. During our research on the use of a stratified charge we
had observed that the range of burning of a petrol-air mixture
could be extended enormously on the weak side by the addition of
a little hydrogen and that without any trace of detonation, so long
as the petrol-air mixture was diluted with so large an excess of
air that ignition from the spark plug would be impossible. In other
words, use could be made of the admission of a little hydrogen
to form a pilot flame capable of igniting an extremely weak
petrol-air mixture, and further that so weak a mixture would not

detonate even at a compression ratio much higher than could be used under normal conditions. With so weak a mean mixture strength the whole cycle temperature would be very much lower, and the thermal efficiency correspondingly greater, but the power output of the engine would, however, be barely one-half of that obtainable under normal working conditions.

Barnes Wallis had concluded that to maintain a uniform buoyancy under normal cruising conditions the consumption of hydrogen should be about one-sixteenth by weight of that of the fuel. We concluded that this ratio of hydrogen, whose calorific value was nearly three times that of the fuel, would be sufficient by itself to take care of the frictional and other losses within the engine and just enable it to run idle on gas alone. We carried out a number of tests both on our variable compression engine and on one of the sleeve valve units and found that so long as the mean mixture strength of the cylinder contents was only about half that needed for complete combustion we would be clear of detonation and could employ a much higher compression ratio than would be possible when running on either gas or liquid fuel alone. Furthermore, it seemed that it might be possible to run on kerosene rather than petrol and thus reduce the fire risk. In this connection I recalled that the Priestman oil engine of the last century had achieved cold starting by the use of compressed air which pulverised the liquid so finely as to dispense with the need for a heated vaporiser, and it seemed to me that we might employ the hydrogen for this purpose. We tried this out on one of our sleeve-valve units, employing a small air compressor both to meter and to compress the hydrogen, while the kerosene was metered and delivered by a plunger pump; at the same time we provided a small supplementary port in the sleeve-valve so timed as to admit the spray during part of the suction stroke. After a good deal of experimental work we got this arrangement to function quite well, and found that even with kerosene we could run without detonation at a compression ratio of 6 to 1, and at a consumption of liquid fuel of just under 0.3 lb. per b.h.p. hour compared with about .44 lb. for the petrol engine alternative, and about .38 lb. expected of the diesel engines for the 'R101'.

On the strength of these test results which extended over about two years, we were asked to prepare a scheme design for a sleeve valve hydrogen-kerosene engine of about 550 h.p. at a speed of 1000 to 1100 r.p.m. and a maximum b.m.e.p. of about 60 lb. per sq. inch. This was a fascinating exercise. Owing to the low mean pressure the total cylinder volume would need to be at least double that of a petrol engine of equal power and speed. On the other hand, the maximum cylinder pressure would only be about 250 to 300 lb. per sq. inch as compared with about 600 lb. in the petrol, and probably nearly 1000 lb. per sq. inch in the diesel engine; thus we could safely employ much lighter scantlings and working parts. Again, the whole cycle temperature would be so much lower that we need hardly worry about thermal stresses, and in short the design problem resembled much more nearly that of a steam than an internal combustion engine. After much deliberation we prepared a scheme for a six-cylinder in-line engine of 10 inches bore x 12 inches stroke. Since with such low pressure friction losses would bulk rather large, we proposed a built-up crankshaft with roller bearings for both the connecting-rod big ends and journal and very light aluminium slipper pistons. For the hydrogen supply we provided a small three-cylinder radial compressor running at crankshaft speeds, and for the fuel supply a gear pump and distributor for metering the fuel to each of the cylinders running at half engine speed. As to the other accessories such as magnetos and oil and water circulating pumps, we followed the usual petrol engine practice.

During the design stage we were in close touch with Barnes Wallis who was always most helpful, and we all came to look upon him as a tower of common-sense, sound judgement and ingenuity. However, as time passed, he became too pre-occupied with other aspects of the problem to pay much attention to a long-range engine development, for it had been decided that the 'R100' would be propelled by Rolls-Royce 'Condor' petrol engines in the first instance, to be replaced later by hydrogen-kerosene engines if and when they proved a success.

We were instructed by the Airship Guarantee Co. to prepare a complete design and detail working drawings of just such an airship engine we had proposed and, at the same time, Brotherhoods were given a contract to build a prototype engine to our design.

We set to work with enthusiasm on what promised to be a thrilling task; little did we realise into what a hornet's nest of politics and intrigue we were entering. What had begun as a technical controversy about the relative merits of the airship and the aeroplane for civil transport, was rapidly developing into a party political conflict as between State ownership and private enterprise. In the meantime, thanks to the more powerful aero-engines developed during the later phases of the First World War which enabled aeroplanes to cover much greater distances or carry much greater payloads than hitherto, the argument for the airship on economic grounds was beginning to wear thin, and its proponents shifted to the argument of greater safety due to the airship's buoyancy and, as some claimed, freedom from fire risk. As to the latter, the myth that the State-owned 'R101' would be inflated with helium instead of hydrogen still persisted in spite of warnings from the technical advisers that its use was out of the question. None the less, ignorant or unscrupulous politicians were claiming that the 'R101' would be immune from fire risk since both its fuel, diesel oil, and its gas were non-inflammable, whereas the 'R100' was to employ both hydrogen and petrol, a notoriously dangerous combination.

With the accent now on safety, it was all the more essential to ensure the absolute reliability of the airship's power plant. To this end it was decreed that in future all the accessories upon whose functioning the reliability of our engine depended must be duplicated and we were given a list of some sixteen accessories, all of which had to be driven from the main engine. To accommodate and provide drives for each and all of these accessories, some of which, such as the centrifugal governor, were manifestly unnecessary, drove our drawing-office almost to distraction. In vain we pleaded that the whole engine was a novel experiment and that the urgent need was first to find out how it would behave, and later decide what accessories would really be needed.

The designing of this engine had monopolised a large part of our drawing-office for nearly two years, by which time both Brotherhoods and ourselves were getting thoroughly exasperated with the whole project. Meanwhile, Chorlton had done a wonderful job with the design and development work of the diesel engines for the 'R101' but was still in trouble with crankshaft torsional oscillation and with failures of his main structure due to very high cylinder pressures, to combat which he had had to provide a much more massive crankcase and cylinder block, and doubts were beginning to creep in as to whether the engine would be light enough and ready in time. In the event they were ready but during the course of development they had put on a good deal of weight, and were considerably heavier than the designers of the airship had allowed for. Considering the state of the art at that date Chorlton had achieved a remarkable performance, for no-one had produced a diesel engine of anywhere near comparable power and weight.

At long last our hydrogen-kerosene engine at Brotherhoods was completed and put on test but without all the duplicated accessories. Rather to my surprise it performed quite well in that it developed the power expected of it but its fuel consumption was rather high. On stripping it down after the test it was obvious that the distribution of fuel between the cylinders was very uneven and that the fuel and hydrogen injectors were not giving a fine enough spray, with the result that a large amount of liquid kerosene had found its way into the crankcase. There were also several other mechanical defects and it was obvious that a lot of patient development work lay ahead.

Thus far we had got when the terrible disaster to the 'R101' brought the reign of the rigid airship to an end. The successful 'R100' which had made a maiden voyage to Montreal and back only a few months earlier was broken up, and all further work on our hydrogen-kerosene engine was abandoned, to our intense relief. What had started as a fascinating technical problem had become so bedevilled with politics that we were thankful to be quit of it.

About 1921 we started carrying out investigations and research directly for the Air Ministry under a research contract which was renewed annually for the next twenty-five years. Under this contract we were requested to investigate all possible means of improving the fuel economy of aero-engines, whether by fuel preparation or by mechanical design.

During the closing stages of the First World War the submarine menace was the greatest danger that our country had to face and it was natural therefore that anti-submarine devices and tactics should play a very dominant part. Above all the Air Ministry were calling for machines with a very long range capable of patrolling far out into the Atlantic and of many hours of endurance; such machines were not required to attain high altitude nor was speed of itself of great importance. Our terms of reference by the Air Ministry left us a very free hand. Briefly they were to investigate all possible means of achieving a really low fuel consumption under cruising conditions. At this time it should be recalled that the octane number of the best aircraft spirit obtained by judicious blending was only about 60. Anything therefore that could make possible a higher ratio of compression would give a very valuable return, for an increase of one ratio of compression in this region would pay substantial dividends.

In the struggle for better fuel economy three possible lines of development were investigated. First there was the sleeve-valve engine, for I had reason to believe that the absence of a hot exhaust valve and the almost ideal combustion chamber form would allow such an engine to work with considerably higher compression and therefore give a better fuel economy. Second was the stratified charge engine which, under cruising conditions should, on paper at least, give a very good fuel economy, and third was the compression ignition or diesel engine. Work on the compression ignition engine was being carried out at Farnborough on a single-cylinder unit and this engine gave extremely good results, thanks to the skilful design of Mr. H. B. Taylor.

Since it appeared at that date that the diesel development was being looked after by Farnborough and in very capable hands, we rather restricted ourselves to the other two alternatives. Experience we had gained with the double sleeve-valve engine of the Knight type used by Daimler with purely reciprocating sleeves had convinced us that this system could not be developed into a highly efficient unit. The alternative, the single sleeve valve with combined rotational and reciprocating movement, as patented some twelve years previously by Burt & McCollum, seemed a much more promising line of attack. It was difficult to get any data as to the behaviour of the Burt type of sleeve under high duty conditions for although it had been marked by Messrs. Argyll and Messrs. Arrol Aster for several years as a very successful car engine this was at a low rating. There were, therefore, many questions to be answered. For example, would the pistons be able to get rid of their heat through a sleeve valve, through an oil film and then through the cylinder walls? Could a sleeve-valve mechanism be designed which would stand up to high pressures and high speeds? What would be the conditions as to combustion and turbulence, and what form of combustion chamber would be suitable? These and many other questions remained to be answered. Our first step, therefore, was to design and construct a single cylinder unit of a size selected after discussion with the Air Ministry authorities.

Of the original inventors of the single sleeve McCollum had died and Burt had long since retired, hence it was difficult to get hold of any reliable technical information. However, once we had achieved a satisfactory design for the operating mechanism of the sleeve valve we were able to carry on with our research on the behaviour of the engine known as the E30/1. This developed into a long story which I will not attempt to deal with here.

In parallel we designed and built a poppet-valve engine of the same dimensions as the E30/1. This engine had four overhead valves with the sparking-plug in the centre, as in the case of the sleeve-valve engine, and during the initial comparative tests the form of the combustion chamber was exactly the same in both engines. In both the compression ratio was, I think, about 4.6 to 1. Fuels blended with heptane and toluene as both pro- and anti-knock dopes, adjusted in each case to promote incipient detonation, and then tested in the E35 variable compression engine showed that the sleeve-valve could operate successfully at one ratio higher compression within the limits set by detonation.

From these tests we reckoned that the sleeve-valve engine should be capable of giving about 10 per cent more torque and at cruising speeds probably about 12 per cent to 15 per cent lower fuel consumption. It was evident, therefore, that the potential advantages of the sleeve-valve engine were so great as to justify a very thorough investigation of its possibilities, and thus started the programme of research and development of the single sleeve-valve engine which eventually resulted in the adoption of the sleeve-valve for all their aero-engines by the Bristol Company, by the Napier Company and, latterly, by Rolls-Royce. The observation that the use of the single sleeve-valve enabled us to operate at one whole ratio of compression higher than similar poppet-valve engines was amply confirmed by numerous tests in later days, and once the potential advantage of the sleeve-valve had been confirmed, the investigation developed largely into one of exploring the many and various mechanical and combustion problems which would be involved when running a sleeve-valve engine at high duty.

It had been my firm belief that the role of our new company would be that of carrying out research, design and development work on behalf of the smaller firms who lacked the necessary resources or experience which we could provide. In this I was utterly wrong and we paid heavily in the early years for my error of judgement.

During the course of the First World War there had come into being a large number of small firms who specialised in the production of individual component parts for aero and other engines. In this field they had served as sub-contractors to the larger firms who were turning out complete engines; some at least of these small firms were well equipped with up-to-date tooling and had done excellent work. With the end of the war their role as sub-contractors came to an end, and they were looking round for new fields of activity. The success of our Tank engine, not only in Tanks but also for industrial use generally, had resulted in several of them applying to us for designs mostly of small industrial engines for which there appeared to be a wide market. With several of such firms we agreed to supply designs and detail working drawings in return for a comparatively small royalty on the sale of their product. Little did I realise at that time that such firms had neither the sales organisation nor the servicing facilities to take care of their production, nor, as sub-contractors, had they had any need for such. Again, as makers of component parts they had no need for skilled fitters to assemble, erect and test their engines. Thus we found ourselves obliged not merely to supply working drawings but also very complete instructions and, in some cases, to supervise manufacture and testing, all of which involved a heavy load on our small staff, but even worse was to follow, for after they had succeeded in selling a few engines they expected us to help in servicing them.

This mistaken policy on my part, combined with our addiction to long-range research, cost us the loss of Halford who had played so valuable a part at the outset of our career as a firm. Neither long-range research nor the design of industrial engines appealed to his dynamic and restless spirit. He therefore decided to set up on his own as a consultant; in this capacity, and with the help of two very able designing draughtsmen, Moult and Brody, he designed a most successful aero-engine, later known as the 'Gipsy' for his friend Geoffrey de Havilland's new light aeroplane the 'Moth', and thereafter he went from strength to strength in his chosen field.

We were all very sorry to lose Halford, but we kept in close touch with him and remained good friends for the rest of his life.

While our association with several small firms had proved disappointing, it was quite otherwise in the case of large firms. I had supposed that the large well-established firms had, during the war years, learned to rely on their own research rather than on the purchase of 'know-how' from abroad, and that therefore they would have little need for our help. I was agreeably surprised to find that though in most respects they were much better equipped with facilities for research than we were, they none the less welcomed and were prepared to pay for our co-operation, partly no doubt because we could serve as an independent check on their own findings, and partly in the hope that we might be able to contribute some new lines of thought. I was delighted, therefore, when such firms as Rolls-Royce, Napier and Bristol in the aero-engine field, Mirrlees, Crossley and Brotherhoods in the industrial, and Vauxhall Motors and Tilling-Stevens in that of motor vehicles entered into consulting agreements with us during the early years of our existence. With them we got on very happily and thereafter we shied off entering into commitments with the smaller firms.

In February 1928 my father died. This to me was the worst bereavement I had ever suffered, for since my earliest childhood days he and I had always been good companions; though our individual tastes differed widely we shared the same philosophy of life and had much in common. My father had always taken the line that of the capital he possessed, my share would take the form of the best possible education he could provide for me, while the rest he would leave to my two sisters, for whose education there was, in those days, less opportunity. It must have been a disappointment to him that in spite of his expenditure on my education, my academic career both at school and at the university had been utterly undistinguished, though he never lost hope that I would eventually make good. In his later years it had become his ambition that one day I might be elected to the Fellowship of the Royal Society. I pointed out that I had no claim to such a distinction, that although I had achieved some success as a mechanic, as a scientist I had no suitable qualifications. None the less he clung to this hope till the date of his death.

It was barely a year after my father's death that I received one evening a mysterious telegram which read 'my heartiest congratulations – Smith'. I was completely puzzled by this for I could think of nothing on which I was to be congratulated, nor did the surname of the signatory provide any clue until I suddenly remembered that Sir Frank Smith, F.R.S. was secretary of the Royal Society, and I wondered. I did not have to wonder long for next morning I received a shower of congratulatory telegrams from Fellows of the Royal Society on my election. This honour came as a complete surprise and rather took my breath away, for although in my secret heart I had cherished the ambition that one day in the far distant future I might achieve it, I had never had the slightest inkling that it would come so soon. My one regret was that my father had not lived to see the realisation of his hopes.

Harry Ricardo with Harry Horning onboard ship, 1924
By permission of Ricardo Consulting Engineers

The Ricardo turbulent combustion chamber
By permission of Ricardo Consulting Engineers

CHAPTER 15

Patents and Royalties

Long before the outbreak of the First World War I had applied for and been granted several British patents relating to the design of my 'Dolphin' two-stroke engine, and also two patents, one on the detail design of the crosshead piston and the other on the use of the masked inlet valve for my design of supercharged aero-engine. Both these patents were embodied in the engines for Tanks.

Although the use of the crosshead piston had contributed so much to the success of these engines, its application increased both the cost and the height of the engine to an extent which would rule it out of the automobile market. In designing a piston for such engines I aimed both to retain some of the advantages of the crosshead piston and, by better distribution of the material, to provide a considerably lighter piston than the conventional trunk type.

After the formation of our company we applied for and were granted patents on this form of piston not only in this country but also in the more industrialised countries of Europe and America, and within the next year or two it came into wide use, particularly in France, Germany and Italy. For this, which generally became known as the 'slipper piston' we asked only a small royalty, for we were feeling our way in the handling of patent rights. None the less, royalties on our slipper piston alone amounted to well over £2000 per annum derived for the most part from specialist firms of whom the German firm of Mahle took the lead.

I have already described how Evans and I while still at Walton arrived at what became known as the 'turbulent head' for side-valve engines. On this also we were granted patents in all the leading European countries, but the U.S. Patent Office put up one objection after another on the grounds of prior art. Their citations appeared to be frivolous or irrelevant but after a long struggle we abandoned the attempt to obtain an American patent in our own name.

During the course of our research on fuels for petrol engines we had come into contact with Mr. Harry Horning, an American citizen and President of the Waukesha Motor Co. who, at that time, were manufacturing side-valve petrol engines. Horning, like myself, was deeply interested in the problem of matching the fuel to the engine, and we became great friends and remained so until the day of his death. To him we demonstrated our variable compression engine and our turbulent cylinder head and told him of our vain attempt to obtain an American patent on the latter which, at that time, had been adopted by several motor-car firms in England, France and Germany.

Horning suggested that if we cared to give him a free hand he felt sure that as an influential American citizen he would have little difficulty in obtaining the patent that had been denied to us. In return we agreed to a purely nominal royalty on his part, and he did in fact obtain a patent.

It was not until the middle 20's that the turbulent head really came into its own and yielded royalties ranging from £10,000 to over £20,000 per annum, and thus formed our main source of revenue during the lean years of the late 20's and early 30's. While our British and Continental patents were respected and royalties were paid for their use, this was not the case in America where several of the leading manufacturers of automobile engines employed our turbulent head but declined to acknowledge or pay royalties on the U.S. patent nor was Horning, at that time, prepared to fight a patent action against such formidable opponents, and there was nothing we could do about it.

By the end of the 20's a considerable number of American cars equipped with our turbulent head were being imported into this country. The Americans were no doubt fearful lest we should levy royalties on their imported vehicles (as legally we had the right to do) for from our faithful licensees in this country and France we learned that American agents were inciting them to challenge the validity of our patent and offering to supply evidence which, as one spokesman put it, would 'bust our patent wide open', and thereby relieve them of the obligation to pay any further royalties. Thus we were faced with the probability that sooner or later, one or more of our licensees, backed by American support, would force us to take legal action in defence of our British and European patents, as indeed happened in the years that followed.

By about 1930 our research and development work on the high-speed diesel engine was beginning to bear fruit. We had already been granted several diesel patents on which royalties were beginning to come in and upon which we relied to replace those from our turbulent head patent whose life was nearing its close. It was all the more important, therefore, for us to make it clear that we were prepared to fight for the sanctity of our patents, present or future.

We began to suspect that certain British and Continental firms also were disregarding our patent and introducing new models of side-valve engines with turbulent heads without payment of royalty. Since there could be no external evidence of this we asked a number of local garages to let us know when they were removing the cylinder heads of their customers' cars so that we could take an imprint or plaster-cast of their combustion chamber shape. By this means we discovered that several firms were in fact employing a form of combustion chamber which we considered fell within the claims of our 1919 patent. One in particular, namely, the then new Hillman 'Wizard' car engine appeared to be almost an exact replica of that described in our patent application. We wrote a polite letter to the Humber-Hillman firm pointing this out, to which we had a very terse reply to the effect that they had been advised that our patent was invalid, and that they were not prepared to pay any royalty. This was a direct challenge and my Board were unanimous that we must take it up, cost what it might.

Although on previous occasions I had been asked to act as an expert witness in other people's patent actions I had always declined such invitations and was abysmally ignorant of the conduct of such cases. With the prospect of being involved in one in the near future I attended and listened in to several cases in the Law Courts to hear just how they were conducted and to study the technique and personality of the leading technical barristers and expert witnesses of the day. I was tremendously impressed by the painstaking care which the judge took thoroughly to acquaint himself with the circumstances of the case however technical it might be, and by his scrupulous care to protect technical witnesses from being entrapped or confused while under cross-examination. In short, I was favourably impressed by the patience, fairness and dignity displayed by all concerned.

For our case against Humber-Hillman we had been advised to enlist the firm of Bristow, Cooke & Carpmael to act as our solicitors and to brief counsel on our behalf. George Cooke, son of the senior partner of the firm, took immense pains to collect all the evidence we required, some of it in the form of affidavits from large-scale users of our patent, such as the London General Omnibus Co. and several others.

On Cooke's advice we retained Sir Stafford Cripps, K.C., as our leading counsel. Cripps warned us that his time was so fully booked that he doubted whether he would be able to attend throughout the whole of the case and urged that we should retain also Mr. Whitehead, K.C., as well as his junior, Lionel (later Sir Lionel) Heald, Q.C., to act as deputies when he himself was unable to attend. I had not seen Whitehead in action but Cooke assured me that he was a first-rate advocate and a pastmaster at diplomacy.

Months passed before a date could be fixed for the hearing of our case and during that time I had several meetings with Cripps who told me that he had decided to call upon me to give evidence in the case. It was, he said, very unusual for an inventor to be asked to give evidence about his own invention lest he overstate his case or get rattled under cross-examination, but he thought that I could be trusted to do neither, while the fact that I was a Fellow of the Royal Society would command respect. He told me that the opposition would claim (1) that my invention had been anticipated, (2) that the royalty we demanded was excessive, and (3) that the claims set forth in our patent application were not sufficiently precise. As to the first two of these pleas, Cripps considered that thanks to George Cooke's briefing he would have no difficulty in disposing of them, but that the last exposed the weak places in our armour, and that much would depend on the interpretation the judge chose to put upon the claims.

At long last a date in October 1933 was fixed for the hearing of our case before Mr. Justice Farwell. It had been agreed that Whitehead should open the case on our behalf. Our Board, and indeed all our technical staff, were most anxious that there should be no hint of vindictiveness or rancour on our part. Whitehead's opening statement was a masterpiece of lucidity. He explained to the judge that we were not manufacturers but a research team and, as such, depended upon royalties from our patents to finance the research we were carrying out on behalf of industry as a whole, and that our objective in bringing this case to Court was to establish the sanctity of one of our patents and that it was far from our intention to hold up the British motor-car industry to ransom. He then went on to outline the general aspects of the case. He went on to say that he could call witnesses to rebut the pleas put forward by the defendants. His statement was followed by Trevor Watson, K.C., on behalf of the defendants who said that he would bring evidence in support of the various pleas I have mentioned already, and then followed several days of argument, and the hearing of expert and other witnesses. As to the plea of anticipation by prior art, the evidence, as we expected, was all of American origin and consisted for the most part of the same citations that the U.S. Patent Office had produced when denying us a patent in 1919; namely, published designs of early hot-bulb semi-diesel engines in which the combustion chamber communicated with the cylinder through a restricted passage. We had no difficulty whatever in satisfying the judge that these were quite irrelevant in that the combustion conditions obtaining in a semi-diesel engine were entirely different from those in a spark-ignition petrol engine. In this and throughout the whole of the hearing the judge himself took an active part by asking a number of pertinent questions which revealed his thorough grasp of the technical aspects of the case.

As to the plea that the royalty we were demanding was prohibitive and out of all proportion to the advantage we claimed for its use, the Chief Engineer of the London General Omnibus Co. deposed that several years previously the whole of his large fleet of buses had been re-equipped with turbulent cylinder heads of our design and his records showed that their use had effected a saving in fuel cost of well over £100,000 per annum while, at the same time, the increase in power output, and therefore improved accelleration, had enabled his company to speed up the services. Asked by the judge how much his firm had paid to us in royalties the reply was 'About £5000'. Similar evidence was afforded by an affidavit by Reginald Tilling, then Chairman of the firm of Thomas Tilling, who operated large fleets of buses both in London and in several large provincial towns. Cooke had collected several other affidavits from users of our patent, but the judge said that he was satisfied and did not want to hear any further evidence on that score.

Next came the legal battle between Cripps and Trevor Watson. The latter maintained that according to established custom and precedent the scope of a patent was limited to that defined in the summary of claims at the end of the document, and that in the case of our patent the claims were so loosely defined that they could be interpreted to include almost any form of combustion chamber applicable to a side-valve engine, and that therefore if the validity of our patent were upheld we might claim a royalty on every side-valve engine. Cripps argued that the preamble to our patent explained very clearly what was intended and that, whether in accordance with precedent or not, the claim should be read in the context of the explanatory preamble.

It was at this stage in the proceedings that Cripps called upon me to give evidence. During the course of his examination he traced the whole of my career from my Cambridge days to my election to the Fellowship of the Royal Society and made me recount the evolution of the turbulent head with Evans' help in the winter of 1918–19; how in my workshop at Walton-on-Thames I had carried on with the research that Hopkinson had initiated and how I had investigated the phenomenon of detonation and had satisfied myself that the poor performance of the conventional side-valve engine was due, in part, to insufficient turbulence and in part to its proneness to detonation. During the winter of 1918–19 Evans and I had evolved a form of combustion chamber design both to increase turbulence and to reduce the tendency to detonate, the former by introducing a restricted passage between the combustion chamber and the cylinder in order to create additional turbulence during the compression stroke. We had found that too much turbulence was even more objectionable than too little. We had found during our experiments on a certain typical side-valve engine that the optimum performance was obtained when the area of the restriction in the cylinder head was substantially equal to that through the inlet valve port. We exhibited several plaster-casts of combustion chambers both of the conventional flat pancake form and of the turbulent form as in use by several of our licensees; some of these had been sectioned to show clearly the dimensions of the restriction, as also casts of the inlet ports. In the particular case of the Hillman engine it was apparent that these were almost exactly equal. We exhibited also a plaster-cast of the combustion chamber of the conventional flat pancake type taken from an earlier engine marketed by the same firm. The contrast was obvious.

For the whole of the next two days I was under cross-examination by Trevor Watson throughout which his questions were invariably scrupulously fair and to the point. It resembled a high-level debate conducted in an atmosphere of friendliness and good humour, at the end of which he paid me some pretty compliments, which the judge endorsed, and thanked me for the patience and courtesy I had shown throughout this cross-examination.

In his summing-up and verdict the next day the judge said in effect that if his judgement was to be in strict accordance with the letter of the law he would have no choice but to declare our patent void because the scope, as defined in the statement of claims at the end of the document, was too vague and too wide, as the defendants contended, but that in all such cases the judge had considerable latitude to use his own discretion. He considered that if the claims were read in the context of the preamble to the patent a different interpretation would be put upon the document as a whole. He went on to say that in this case all the evidence he had heard had satisfied him that at the time of application the invention was novel; that its utility had been amply confirmed by such evidence as that given by large fleet users and, in short, that it had been proved a valuable invention arrived at by long and painstaking research. His verdict was therefore that the validity of my patent should be upheld. In giving this verdict he was fully aware that he was departing from established practice and precedent and that a higher Court might have different views. He therefore gave the defendants leave to lodge Notice of Appeal within a period of, I think, fourteen days. As to the question of infringement by the defendant Company the evidence had convinced him that this was established and he ordered the defendant Company to pay the costs of the case and any arrears of royalty due to us. Thus ended a case which had occupied the Court for nine whole days but whose preparation had occupied almost all Hetherington's, Evans' and much of my time and thoughts for a year and more and had cost us over £20,000, and many months of acute anxiety, whereas the statutory costs allowed us amounted, so far as I can remember, to barely one-third of this sum.

We had to keep our fingers crossed until the time allowed the defendants to give formal Notice of Appeal had elapsed, but they did not appeal and the judge's verdict became final. We had made it clear to all that we were prepared to fight and to fight hard for the sanctity of our patent. It now remained for us to let it be known that we had no intention of exploiting our victory by demanding excessive royalties on any of our patents present or future; that all we looked for was to recover the actual costs of our patent action and that without recrimination. To this end we obtained an order from the Court for the payment of arrears of royalty from the agents of those American firms whom we knew to be infringing Horning's American patent and importing them into this country and, at the same time, had been inciting British firms, such as Humber-Hillman, to infringe.

So ended a case the outcome of which might well have meant life or death to our hopes for the future and which had hung over our heads for nearly two years. Before the hearing of this case our Chairman, Campbell Swinton, to whom we owed so much, had died; Goodenough had retired and I had become Chairman of the Board. With the loss of the two most active Board members I felt acutely the need for a wise and experienced counsellor, and was fortunate indeed in getting Kidner to join us. Kidner had retired from Vauxhall Motors when that firm had been taken over by General Motors of America and he became a very active member and, a little later, President of the Institution of Automobile Engineers. He was well known to and held in great respect by all the leading members of that industry for his ability, his wisdom and his obvious integrity. To me he proved to be a tower of strength and encouragement during that critical period. To him fell the lot of making it clear to the industry as a whole that we had fought this case in order to defend the sanctity of our patent and that we had no intention of exploiting the advantage we had gained to demand damages or arrears of royalty from those firms who, acting under the misapprehension as to the validity of our patent had in fact infringed it. He was as anxious as I was that no lingering ghost of bitterness or greed should remain to haunt our future relations with industry, and in this he was, I think, entirely successful, for never again has the validity of any of our patents been challenged.

After the successful conclusion of our patent action only a few months remained before the expiry of our patent and before the reign of the side-valve engine was brought to a close. In this form of engine the position of the valves renders it virtually impossible to employ a compression ratio higher than about $5\frac{1}{2}$ to 1 without restricting the breathing capacity or overdoing the turbulence. By 1934 thanks partly to the discovery by Midgley in America in the early 20's of the remarkable effect of the addition of tetra-ethyl-lead in reducing the tendency of the fuel to detonate, and partly to improvements in refining and processing, the octane number of commercial petrol had been so far improved as to allow of the use in small cylinders of considerably higher compression ratios possible only with overhead valves; thus once again the overhead valve engine scored an advantage over the side-valve even though the latter was equipped with our turbulent head.

Up to this period our royalty charges had been based on an estimate of the practical value of our individual patent. So long as only two or three patents were involved, this rather haphazard assessment had sufficed, but as our research and development work on the high-speed diesel engine proceeded we had taken out a considerable number of new patents. Kidner pointed out that this would inevitably lead to confusion in the years to come and proposed that we should ask our licensees for an omnibus royalty to cover the use of any or all of our patents, present or future, and that such a royalty should be based on some simple dimensional factor, such as the piston area or the cylinder capacity of our clients' engines. This policy we adopted and have not regretted it.

With the successful conclusion of our patent action, and the rationalisation of our policy in regard to patents and royalties and with the assurance of goodwill on all sides, we were able once again to concentrate on our legitimate work on research, design and development which has proceeded smoothly and happily and with increasing volume ever since.

I will not attempt to go any further into the technical side of our work at Shoreham for most of it has been recorded both in the various editions of my book, in the Proceedings of such Institutions as that of the Mechanical Engineers, the Automobile Engineers, Royal Aeronautical Society, etc., while I am almost ashamed of the number of lectures and papers that I and other technical members of our firm have delivered during the years that followed.

Looking back over the first fifteen years of our firm's existence I think that I found the most fruitful lines of pure research that we carried out during our early years at Shoreham were those related to fuels, lubricants and lubrication and to the process of combustion in both spark-ignition and diesel engines; the most fruitless those of the use of the stratified charge on which, in earlier years, I had pinned such high hopes; on what we termed the regenerator engine, a form of low pressure compound diesel engine which proved a complete failure, and our hydrogen-kerosene engine for airships, all three of which were started with high hopes. As to development work, that on the light high-speed diesel engine for road work, and on the sleeve-valve engine for high-powered aircraft proved the most rewarding, while our attempts to develop a diesel engine light enough for aircraft propulsion though it taught us much showed that such an engine could not compete with a comparable petrol engine running on the high octane fuel available for aircraft by the middle 30's. Among other failures for which I was responsible was an attempt to make an open-cycle hot-air engine to fulfil a need by the War Office for a completely silent engine of small power. As to design work, from the very start our drawing office was always fully occupied with the preparation of designs of engines for our clients both in this country and abroad, and for experimental engines for our own use. There were, however, some fascinating jobs; to me the most fascinating of all was a request by Vauxhall Motors early in 1920 to design for them a three-litre racing-car engine for which we had a perfectly free hand. Pomeroy, whom I so much admired, left Vauxhalls to go to America to take charge of the design and development of an all-aluminium car, and his right-hand man, King, a great authority on chassis design, had always left the engine to Pomeroy. King it seemed had welcomed Kidner's suggestion that the design of this new engine should be our responsibility. We completed the design in very short time, as did King that of the chassis, and the complete car was ready for the road in little over eighteen months and in time for me to include a full description and an analysis of the performance of its engine in the first edition of my book published by Blackie in 1923.

Another highlight of the early 20's was the design of a high-powered motor-cycle engine for the Triumph Co. which became a popular favourite for the next ten years or more, but the bulk of our design work was of more conventional engines for heavy-duty road work.

To those of my readers carrying out research of the kind I have been involved in, I would like to offer a few words of advice and a few warnings. First and foremost make up your mind what to go for, that is to say what, in your judgement, will be likely to fulfil a need in say four or five years' time; having once decided keep that objective always clearly in view. Do not let yourself be cast down by disappointment, too elated by those initial successes which so often prove to be only transitory. Do not be afraid of failures. In my experience one learns as much, or possibly more, from one's failures, and I have been responsible for many such, as from one's successes; the downright failure is always instructive and is usually fairly early apparent before it has cost an undue amount of time or money. The real danger, and by far the most difficult to cope with, is the partial success, the achievement which is either not quite good enough, or for which the need is passing. To cope with this taxes one's judgement to the limit; it requires all one's strength of mind to break off when cool judgement counsels the abandonment of a project to which one has grown very attached, and on which one has lavished many months of thought and painstaking research. Such decisions have sometimes to be made and, speaking for myself, I find it easiest when the demon of doubt becomes insistent to suspend all work and thought on the project for a few days or weeks, and then review it afresh. It is surprising how coldly and dispassionately one can review and, if necessary, reject one's own most cherished schemes after they have been banished from one's mind for a decent interval.

Woodside - the house my father built

On The Pearl, c. 1928
Kate Ricardo, David Pearce, Angela Ricardo

Beatrice and Harry Ricardo, 1967

CHAPTER 16

Some Frivolous Pursuits

Although I may have given the impression that my adult life was all work and no play this, except during the war period, was far from being the case. From the time I left Cambridge until I got married I lived with my parents at Bedford Square, though half my time was spent on tours of inspection both in this country and on the Continent. My work as an inspector was not arduous, nor did it involve any homework, so that I had plenty of spare time. I used to spend week-ends either at Rickettswood, where I went shooting, or with my parents at Graffham, where I helped my father with his garden, and went for long walks on the Sussex Downs, which I thoroughly enjoyed.

As to my evenings in London, some of these I spent with my friend Hetherington scheming out designs for new engines, some in my small basement workshop and some others attending lectures, but most of my evenings I spent in much more frivolous pursuits. On Monday evenings my father, who was a very keen musician, used to go regularly to concerts at the Queen's Hall, generally known as 'Monday Pops', while my mother, my sister Anna and I generally went to one or other of the so-called variety theatres, such as the Alhambra, the Holborn, the then newly built Palace Theatre and others, where we saw such famous comedians and singers as Dan Leno, Marie Lloyd, Harry Lauder, Pellisier with his troupe of Follies and other great artistes of the music-halls. On other evenings we would all go to the Gilbert and Sullivan operas at the Savoy Theatre, the music of which I could understand and enjoy to the full.

Those Edwardian years saw the heyday of musical comedy such as *The Merry Widow, The Chocolate Soldier,* and *The Geisha,* and others, with their lovely lilting and sensuous music by such composers as Strauss and Franz Lehar. I enjoyed these enormously largely because of the beauty and charm of the lead-

243

ing lady and the chorus of lovely young women chosen for their beauty and elegance. I went, I think, to every one of these to such theatres as the Gaiety, and invariably fell in love with the leading lady, with the result that my mantelpiece was adorned with post-card photographs of Ellaline Terriss, Lily Elsie, Marie Lohr, Zena Dare, Gertie Miller and other famous stars.

Although we all lived in the same house I saw little of my two young sisters during this Edwardian period. Anna at the time was working at the Slade School of Art in Gower Street. She had just 'come out' and was revelling in the freedom and independence of her newly acquired status. She was generally surrounded by groups of budding geniuses, both male and female, who formed a sort of mutual admiration society, took themselves very seriously, and talked a strange language that was Greek to me. She was for ever going to Bohemian parties or to dances, to some of which she dragged me merely as a make-weight to balance the sexes. I was a clumsy dancer and invariably trod on my partner's toes.

Anna wore her heart on her sleeve and was frequently falling in love with one or other of her many suitors. She did not take herself, her heart or her emotions seriously, and was always prepared to laugh about them, for she had a keen sense of humour.

My sister, Esther, nearly nine years younger than I, was still a schoolgirl during this period. She, too, had her own circle of young friends, all great animal lovers who, when not at school or in the country, spent much of their spare time at the Zoo. She, like my father, loved the country, whereas both Anna and my mother were Londoners at heart.

As time went on one after another of my school and Cambridge friends got married and set up homes of their own, and I longed to do likewise. In an earlier chapter I have recorded my first meeting with my future wife, and that we were married in 1911. Shortly after the birth of our eldest daughter, we went to live in a small house at Walton-on-Thames, where we had about one-third of an acre of garden, and where I at once set up a workshop, large enough to carry out the research I had been longing to undertake. There we spent probably the happiest two years of my life. My firm had moved into new and more spacious offices in Dartmouth Street, Westminster, where I had a room and drawing office to myself, with virtually a free hand to design such mechanical equipment as we needed. I went daily to town by road or train and I had plenty of spare time and all my week-ends to enjoy a new way of life with an equally emancipated wife, a home of our own, a small but manageable garden and, to crown all, a baby daughter of our own. I had always been charmed and fascinated by babies and small children, and had longed for the time when I might have one or more of my own to play with; to watch and be amused by her first attempts at walking, the first coherent speech and her rapidly developing sense of humour.

My wife, like my sister Anna, had been an art student at the Slade. She was, and still is, a very clever artist but neither she nor her circle of friends clothed their art in so much solemnity as did Anna's entourage.

During those months before the First World War burst upon us, our daughter, our house, our garden and my workshop formed our little self-contained world, and we asked for nothing more. We had, of course, plenty of visitors from among both my wife's friends and mine. The latter, I am afraid, I dragooned into helping me in my workshop or garden for, from my earliest schooldays, I had cultivated the art of getting others to do my work for me. Then came the war and, with it, the end of that happy carefree manner of life.

In the preceding chapters I have related my experiences during the war years. Suffice it to say that when the horror of war was over, I emerged a relatively rich man, for not only had my salary been increased greatly, but immediately after the Armistice I received a personal award for my design of engines for Tanks which, after various disbursements, amounted to about £12,000. With our new-found wealth we bought a large, ugly, but very well-built and comfortable country house called Penstone, at Lancing, about three-quarters of a mile from the sea and barely two miles from the new laboratory. I had bought also a second-hand 1911 Alphonso model Hispano-Suiza sports car for my own use, and a 12 h.p. 1912 Humber car for my wife. My old Dolphin car I handed over to my colleague, Evans, who used it for several more years. Thus, with the nightmare of war behind us, and a second daughter born in 1915, we started out in the spring of 1919 on a new era of family life, this time on a more ambitious scale.

During our time at Walton we had become keen on gardening and our garden had been small enough for us to manage alone. My wife looked after the flowers, while I dealt with the fruit and vegetables and we found it fascinating to watch their growth day by day and to experiment with different treatments. At Lancing, however, our garden was much too large to handle ourselves, and we engaged the full-time services of an expert gardener. Almost everything flourished under his skilful craftsmanship. My wife and I learnt a good deal from him but we also suffered, for his was a dominating personality and he soon became tyrannical and would not allow either of us to do any gardening ourselves. In all other respects he was co-operative and helpful, but woe betide either of us if we dared to interfere or criticise his garden.

Penstone and its garden became our Lancing home for the next thirteen years. The house had originally been built by a South African millionaire who had evidently spared no expense on its construction; its thick stone walls kept us warm in winter and cool in summer. It was equipped with central heating and electric light from its own power plant. In short it had all the amenities, and during the years we lived there cost us almost nothing in repairs or maintenance. My father was shocked at its ugliness but thoroughly approved of its internal design and excellent workmanship.

We completed the purchase (just under £2000 all told) and moved in in the spring of 1919. Our two daughters, Kate and Angela, were then aged seven and four years. They were delighted by the spaciousness of their new home, and the nearness of the sea and its sandy beach. My friend and colleague, Oliver Thornycroft, had acquired a house on the sea-front at Lancing. He had married shortly before us and had three children who shared our playroom. Jointly with the Thornycrofts we engaged a young local woman as a nursery governess, an arrangement that worked very well.

During the first few weeks of our residence at Lancing, Thornycroft and I were busily engaged in putting the finishing touches to the new laboratory. Tizard and Pye then joined us, Tizard as our guests at Penstone and Pye with his old friends the Thornycrofts, and we all settled down to what proved to be the most interesting and, I think, the most fruitful research with which I have ever been concerned. At the same time I returned to the sort of home life my wife and I had so much enjoyed immediately before the outbreak of war, and I took up once again the hobbies – boating, fishing and butterfly collecting – of our earlier days. With the work I loved best in the world only five minutes drive away, with more spacious surroundings, with the sea close at hand, and with more money to spend and with more leisure time to enjoy my wife's and my children's company I had everything that the heart of man could desire, except that we never again recaptured the sense of security or the carefree lightheartedness of our pre-war days. Although the politicians and press assured us that never again would there be another major war, yet in our hearts we knew this to be wishful thinking.

In my childhood I used to go sailing and fishing with my father off the Devon and Cornish coasts which he and I both enjoyed enormously. At Shoreham there was plenty of room for small boat sailing inside the harbour at high tide, but I preferred the open sea both for its sense of freedom and for its fish; owing, however, to the swift tidal stream, it was almost impossible for a small sailing boat to enter or leave the harbour except at periods of slack water. As soon as we were established at Shoreham we searched for and found a dinghy equipped with a little single-cylinder, four-cycle petrol engine which, because of its shallow draught and light weight, could stem the stream through the harbour entrance at most states of the tide. Tizard, like myself, enjoyed boating and was a keen fisherman, and we had great fun, but little luck, together at week-ends or in the long summer evenings fishing for mackerel. We therefore decided to try our hands at trawling. We bought a trawl net with a beam about five feet wide, small enough and light enough for one of us to handle, while the other looked after the boat. It took us rather a long time to learn how to handle our trawl net to the best advantage, the chief problem being how to keep the net on the sea-bottom except in very shallow water. With the drag rope heavily weighted, the whole affair became too heavy to lift on board. The alternative of using a very long tow line involved too much time and effort spent in paying out and hauling in the net. We had therefore to compromise between these extremes and to be content to trawl at a maximum depth of about seven or eight feet. Another problem was that the amount of sea-weed collected in the net took a long time to clear. On very calm days, when the light was right and the water clear, and we could see the sea bottom, we could manœuvre the boat to avoid the patches of weed. In the course of time we grew quite skilled in the handling of our small trawl net and often caught quite a lot of skate, plaice and Dover sole, all of small size but large enough for the table. Speculation about each haul lent excitement to our fishing. We usually brought up two or three fish large enough to keep, about double that number of smaller fish which we threw back, and in addition we often caught a small octopus, a squid or a large spider-crab. On good days we could reckon on catching 1 lb. of edible fish per pint of petrol used by our little engine. For these fishing trips Tizard was the ideal companion, and I think he enjoyed them as much as I did; indeed he was the perfect guest during his stay with us in the summer and autumn of 1919. His wit charmed us all, and my two young daughters adored him.

We had had such fun with our motor-boat, which we called *The Mudfish,* during our first two summers at Shoreham that we commissioned a Shoreham boat-builder to build us an eighteen-foot motor launch, and later two small sailing dinghies for our two elder daughters who had also taken to boating.

Our new motor-boat, which we christened the *San Jose,* proved a great success; her little four-cylinder engine was very quiet and smooth-running and she had ample seating capacity for all our family and several friends as well. Beautifully built of well-seasoned timber she served us well for nearly twenty years, as did the children's sailing dinghies. At high tide, the wide expanse of smooth water in the harbour and river estuary provided a safe and sheltered area for them to practise sailing, and they became quite adept at handling their little boats. On summer evenings, when the tide was right, it was our delight to go for supper picnics up the River Adur, which was navigable for about six miles inland, and to watch the bird life on the river banks. At week-ends, when the weather was favourable, we sometimes made the open sea passage to Littlehampton, and thence up the River Arun to Arundel or Amberley. Gradually we grew more adventurous and we longed to have a much larger boat in which we could all sleep on board and make longer passages.

From an old school friend of mine I heard of a boat for sale at Brixham. In her youth she had been a Bristol pilot-cutter, accustomed to making long sea passages in all weathers far out into the Atlantic. In her old age, she had been converted into a trawler, for which service her owner had installed amidships a four-cylinder Thornycroft oil engine of about 30 h.p., driving through a reduction gear a very large propellor. Her name was *Pearl;* her age uncertain but believed to be about sixty years; her length was 48 feet, her beam 13 feet 6 inches, and her draught nearly 8 feet. She was built throughout of oak and appeared to be in good condition despite her age. We fell for her at once and bought her for £400. Meanwhile we amused ourselves planning her internal arrangements and reckoned that we could sleep six in reasonable comfort and eight at a pinch.

It needed only a few trial trips to convince us that we could never tolerate the mess and smell of her kerosene engine. As a replacement we fitted a small 2-litre petrol engine which we had recently designed for a French firm. This I had mounted in the counter immediately abaft the cockpit, from whence it drove the propeller shaft below by chain with a total speed reduction of about 4 to 1. This arrangement, which occupied no otherwise useful space, proved quite successful. We rigged up temporary partitions to form three two-berth cabins, in the largest of which there was just room to take two extra cots.

By the spring of 1924 *Pearl* was in full commission, not a luxury yacht but quite adequate for our needs. It had been our intention to do all our voyaging under sail alone, and to use our auxiliary engine only for manœuvring in or out of the harbour, but since during the summer months the prevailing wind was generally in the west or south-west, and our objective the deep water harbours to the westward, it meant beating all the way and this, with our heavy sails and running gear, was rather exhausting. We found, however, that with just a little help from our engine we could sail much closer into the wind. It also helped us to go about at the end of our tack, hence we got into the habit of keeping our engine running light all the time when contending with a head wind. *Pearl's* deep keel was a great asset when sailing, but an infernal nuisance in that it forbade us the use of any harbour in which she could not lie afloat at all states of the tide. It became our practice, therefore, to keep her in Poole Harbour during the summer months and from there to cross to the westward to such harbours as Dartmouth, Plymouth and Fowey. We kept *Pearl* for about twelve years, but old age was telling on her. Her keel fastenings were threatening to give way and she was no longer seaworthy, so eventually we sold her to a boat-builder in Rye who converted her into a houseboat. During her life she gave us a great deal of pleasure and the thrill of adventure. So far as I can remember we never got into any dangerous situations for she was an excellent sea boat, and seemed to enjoy rough weather, but, being all amateurs, we did get into several humiliating and undignified positions, a few of whichstand out in my memory. For instance, to watch the Schneider Trophy Seaplane Race of, I think, 1928, the Admiralty had allotted us an anchorage on the edge of the course near Ryde, an ideal position, but the tidal stream was running very strong at that point, our anchor would not hold and, during the race, we kept drifting on to the course. The whole area was so densely packed with shipping of all sizes that there was no room for manœuvre, and we got severely ticked off by a naval patrol boat. Worse humiliation was to follow, for when the race was over we met a sister ship to the *Pearl* and started to race her. Seeking to steal a march on her I took the risk of cutting just inside the buoy marking the end of Ryde Sands, and ran hard aground. The tide was then falling rapidly; the *Pearl* slowly heeled over and lay on her side and became the laughing-stock of the assembled shipping, all the more so since she was lying on the wrong side of the marker buoy, and there she lay until after midnight, the object of ridicule and shame. The next and final Schneider Trophy Race we watched from on-shore!

Both my sisters got married soon after I did and produced between them nine nephews and nieces; meanwhile my wife's two brothers contributed five more and in 1921 our third daughter, Camilla, was born. Like Rickettswood in my grandfather's day, Penstone too, during the summer months, was well populated with daughters, nephews and nieces and we were thankful for our large garden and play-room. Fashions had changed since my early days at Rickettswood, and children were no longer segregated to the nursery, but were to be heard as well as seen. We much enjoyed these large school-room parties, whose average age at first was about four or five and, above all, it was wonderful, after a lapse of over five years once again to have a baby of our own sprawling on the hearth rug, for the worst thing about children is their tiresome habit of growing up. At the same time the house was filled with my wife's Women's Institute and other activities.

During the summer of 1919 Tizard had stayed with us as a most welcome guest, but during the next two summers he and his family rented a house on the outskirts of Worthing. During our first few years at Penstone we spent our summer evenings boating and fishing, playing tennis and playing games with the children. During the winter evenings I spent most of my time, after the children had gone to bed, writing the second volume of my book dealing with high-speed internal combustion engines. That it had lain dormant throughout the war was really an advantage in that I had gained far more experience and could write with first-hand knowledge of the research and experimental work we were carrying out daily at Shoreham. This was finally completed in 1922 and published in the following year, together with Volume I which had lain in proof form since 1915, and was sadly out of date. There was nothing I could do about it beyond making a few minor corrections. Other evenings I spent in my spacious workshop, sometimes making toys for the children, such as model boats or strange wheeled vehicles, and others making small pieces of equipment for our research.

Penstone stood at the northern end of the village of Lancing. Beyond it to west and north lay a stretch of open country, while here and there were a few small farmhouses, and a mile away rose the Sussex Downs in all their unspoiled grandeur. The whole of this area of coastal plain had formed part of a large private estate, but owing to death duties it was being sold off piecemeal to speculative builders, and before we left in 1933 it had become a completely built-up area, with rows of shops, street lighting and the buzz of traffic. Penstone and its garden had become hemmed in on all sides by a vast new building estate which was spreading ever higher up the slopes of the Downs. All the charm of the countryside was fast disappearing, and my wife and I were scouring the neighbourhood in search of a new home.

On the north side of the Downs we had looked longingly at a small but very charming old manor-house nestling close under the Downs in quite unspoilt country and yet only about six miles from my work. Tottington Manor estate comprised about 250 acres, including ninety acres of woodland, 120 acres of farm land and about forty acres of Downland. It included also two cottages for farm-workers and a number of farm buildings. It fronted on to a little-used road which skirted the north side of the Downs.

We had never contemplated becoming landowners on so large a scale, nor did we know anything about farming, but the whole place had an irresistible charm for us. What did appeal to us was the protection a larger area of property afforded against the invasion of bricks and mortar which in only a few years had engulfed Penstone. To the south the property extended to the top of the Downs, while a large area of woodlands formed a protective screen to the north against any invasion from that, the most vulnerable quarter, while, for various reasons, there seemed little risk of encroachment from east or west. We hoped to be able to let the farming to a good farmer, leaving us the ninety acres of woodland as a playground and to form a bird and butterfly sanctuary.

After much heart-searching and bargaining on our behalf by a solicitor, we bought the entire property for the sum of £7000, and let it be known that we were ready to let the farm. Our luck was in for we heard of a young man, the son of a local farmer, who had just finished his training at an agricultural college and was anxious to try his hand at running a farm of about our size, and this he did with great success. We could not have wished for a better tenant, or a more helpful or companionable neighbour. He never grumbled nor made unreasonable demands but was always ready to co-operate. With the help of two experienced farm-hands living in our cottages, he very soon turned the holding into a prosperous enterprise as a dairy farm, which yielded him a handsome and well-deserved profit in the years that followed.

At Tottington Manor we started once again on a new mode of

life in the depths of the countryside. Our house dated from Elizabethan times; it was small with a Horsham stone roof, but subsequent owners had added a new wing, with one large room on the ground floor, and above it two bedrooms and a bathroom. The rooms in the old part of the house were all small and with low ceilings supported by oak beams which sagged ominously, very picturesque but inconvenient. Few of the rooms were truly rectangular or the floors level, which made furnishing difficult. None the less, the old part of the house was cosy and compact and we became very fond of it during the years that followed.

During my nursery and early school age one of my hobbies had been butterfly collecting, and I had amassed a good collection of British butterflies and moths, many of which I had bred from the caterpillar stage. As time went on I lost interest in collecting, but not in admiring them, for their beauty and charm made a great appeal to me. To my delight, our large area of woodland teemed with butterflies, not only the commoner species, but with such beauties as White Admirals, Silver Washed and other Fritillaries and rarities such as Marbled Whites. One of our first undertakings was to make a clearing in the depth of the woods where we could sit and watch the wild life far removed from the sight, sound and smell of engines, and sheltered from all the winds of heaven. In early spring our clearing was carpeted with primroses, anemones and wild violets. Here, too, we planted a few flowering shrubs such as buddleia, and flowers such as sedum, which experience had taught us were the butterflies' favourite delicacies. Thus, they came to make it their playground, and this silent, secluded spot became our favourite haunt for reverie, siesta and picnics during the next eight years.

I became ambitious to populate our area with that rare and beautiful butterfly the Swallow Tail, which had become almost extinct in this country: its favourite food is fennel and of that there was plenty growing wild in and around our woods. We made friends with the keeper of the Insect House at the Zoo and he provided us with clusters of eggs. When the young caterpillars emerged, we spread them among clumps of growing fennel. To protect them from birds, we enclosed the whole clump in a tent of butter-muslin, and left them alone. In this way we reared about thirty or forty Swallow Tail butterflies, most of which were released in our woods, keeping in captivity only three or four pregnant females to provide a new generation. Thus we carried on for several successive years during which time we must have released well over 100 butterflies. We had hoped that they would make the area their home, but in this we were disappointed for they all disappeared and our attempt to populate the countryside with a new species failed completely.

While living at Walton before the First World War my wife and I used to breed a few caterpillars of Red Admirals, Peacocks and Tortoise Shell butterflies. When they reached the chrysalis stage, we attached them to, or suspended them from, a branching twig, thus forming a miniature Christmas tree with perhaps a dozen or more chrysalises attached. This we kept by our bedside to watch the emergence of the butterfly, always in the early morning. About twenty-four hours before emergence, the thin opaque sheath of the chrysalis became semi-transparent and we knew what to expect; soon after sun-rise the chrysalis, which had remained dormant for the past three or four weeks, would start to wriggle convulsively until the thin enclosing sheath split open, and the insect would scramble out, a damp woe-begone object with a long black body, its wings a mere crumpled bundle on either side. Slowly and nervously it would climb along the underside of the twig, there to hang motionless while its limp wings slowly unfurled themselves and hung down like damp linen on a washing-line. In a few minutes, however, they would stiffen and straighten themselves as though they had been ironed. Next, the insect found power to move its wings, and its whole bearing would become more confident. After a few tentative and cautious flaps, it would clamber round to the top of the twig and there, for the first time, spread its wings in all their glory, and one could sense the pride it felt in this achievement. The next step was experimental flight, and after a good deal of wing flapping it would risk loosening its grip on the twig until, for the first time, it became airborne. Then, if one put one's finger close to and a little below the twig, it would screw up its courage and let go with its legs, and plane down on to one's finger where it would stay for a few minutes flapping and testing its wings. Again, one could sense its pride in its new achievement. After a few more experimental flights, each more ambitious than the last, it would fly out of the window in search of a mate.

Our garden at Tottington, though much smaller than that at Penstone, was too big for us to manage alone, but this time we were determined that it should be our own garden to do with as we wished and not ours on sufferance as had been the case at Penstone. From the village of Small Dole, about a mile to the north, came a young man named Waller who offered his services. He was very naive and unsophisticated, but had a charming smile which won our hearts. He claimed no skill or experience but said he would be willing to try his hand at anything in the way of outdoor work. We engaged him on the spot and he remained with us all the time we were at Tottington.

Once again my wife undertook the flowers and I the fruit and vegetables, while Waller relieved us both of the heavy work such as digging, carting away refuse and mowing the lawn. He was always unfailingly cheerful and entered with zest into my experiments in horticulture. One of these was the growing of vegetables under removable glass cloches which were then in fashion. I soon found, however, that the labour involved in constantly removing and replacing each cloche one by one for watering and weeding was far too great for my taste, while the casualty rate in broken glass was excessive. Moreover, it involved a great deal of stooping, a posture which I have always hated. Each individual cloche covered an effective area of little more than one square foot, while the available height was very limited. I concluded, therefore, that what I really needed was a huge cloche covering a large area of ground, one in which I could stand upright, do the necessary weeding with a long-handled hoe and the watering with a hose : in short, a complete, but none the less, movable greenhouse. Movable it must be because as with the cloches I intended to prepare the ground and raise the plants in the open and only to cover them for a couple of months at a time while the plants were maturing or fruit ripening. The problem was how to move my giant cloche from place to place without disturbing the structure and so breaking or at least loosening the many glass panels. After much deliberation I decided that the best method would be by floatation, by which means I could move a very large structure without the least risk of strain or distortion, and I wanted a large structure.

I planned, therefore, to construct a greenhouse 20 feet in width by 30 feet long mounted above two concrete water troughs spaced 20 feet apart from centre to centre and 120 feet long; thus between the troughs I would have four separate plots over any one of which my greenhouse could brood in turn during the appropriate season. To the best of my recollection the internal dimensions of the troughs were 13 inches wide at the top, 12 inches at the bottom and about 12 inches in depth. With the help of Waller, our farmer tenant and Mr. H. Hersey from our works at Shoreham, and following the advice of the manager of the local cement works, we formed the troughs in separate 30 foot lengths, leaving a gap of about $1\frac{1}{2}$ inches to be filled in with a special bitumen mixture, provided to serve as an elastic expansion joint. This worked very well, for in the years that followed we never had the slightest trouble with leakage or cracking.

As to the structure of the greenhouse itself, this was little more than a light wooden framework glazed all over, the roof being 5 feet 6 inches high at the eaves and about 11 feet at the ridge. The roof was well trussed with light steel tie-rods and carried on wooden 'A' brackets pitched about 3 feet apart along each side. The bases of these brackets rested on a straddled beam across the open top of the concrete troughs. The inside limb of each bracket was vertical, and the outside, which carried the glass, sloped inwards towards the eaves; thus, I had inside the structure 30 feet of open water tank along each side. As to the ends of the greenhouse, these also were glazed all over except for the lower 3 feet which I left open to allow of the greenhouse passing over growing plants. Once in position they were closed by readily detachable glazed panels.

To float the greenhouse I made six pontoons each consisting of a very light wooden framework covered with sheet tin, each about 10 feet long, 11 inches wide and 11 inches deep. These were quite light to handle. All one had to do was to submerge them into position three on each side, a very simple matter. I had provided inside the greenhouse two large semi-rotary pumps intended both for watering the plants and for pumping out the pontoons which were connected together by loose syphon pipes. As the pontoons rose their rims engaged with the bases of the 'A' brackets; it required only a few more minutes of pumping for the whole structure to become water-borne. It could then be floated slowly along the whole length of the troughs, and that with the minimum of physical effort. Once we had manœuvred it into position on its new site it was anchored securely to bolts which we had let into the side of the troughs, and the pontoons removed. After a little practice Waller and I found that the whole operation, from start to finish, could be accomplished in less than two hours.

We did not, of course, attempt to float the greenhouse in a high wind, but since this operation was carried out only three times a year, we could always wait for a fine day. With a light breeze in the right direction we could let it sail itself, much to the amusement of our guests.

We named our greenhouse 'The Queen Mary' for she made her maiden trip in the same year as that great ship. In the years that followed she proved very effective; on each of our four available sites she covered an area of over 400 square feet, or the equivalent of over 300 cloches. It became our practice to plant one section with winter vegetables such as lettuce, another with strawberries and raspberries, and a third with tomatoes, leaving the fourth fallow. During the winter months she brooded over the vegetables; in February or early March over strawberries, which she ripened in early May, and then over tomatoes for the rest of the summer. We experimented, of course, with other crops, always keeping one section fallow to allow of rotation. During most of the year our troughs kept full, for the catchment area of her roof was about 600 square feet and drained into the troughs. It was only on rare occasions that we had to top them up with a hose from the farmyard.

I had had great fun over both the design and the construction of 'The Queen Mary' over which my daughters, their boy friends and other visitors took a hand. Compared with cloches, it was horticulture de luxe. In all weathers, with plenty of space, of water and of head room, I could garden in comfort at any time of the year.

Our last two or three years at Tottington were rather sad ones. The war clouds were gathering ominously again. We had no babies or young children to liven us. In July, 1940, after the invasion of France, our house was requisitioned by the military, and we retreated to Oxford where we remained during the Second World War.

Bridge Works, Shoreham-by-Sea, Sussex, photographed in 1990
By permission of Ricardo Consulting Engineers

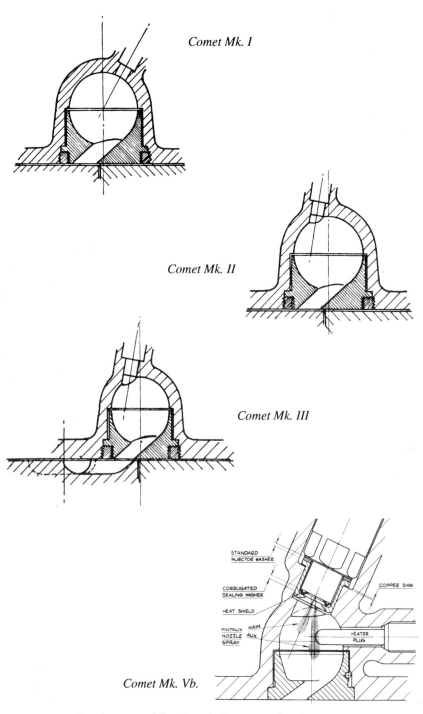

Comet Mk. I

Comet Mk. II

Comet Mk. III

STANDARD
INJECTOR WASHER

CORRUGATED
SEALING WASHER

HEAT SHIELD

PINTAUX   MAIN
NOZZLE   AUX
SPRAY

COPPER SHIM

HEATER
PLUG

Comet Mk. Vb.

Development of the Ricardo Comet combustion system
By permission of Ricardo Consulting Engineers

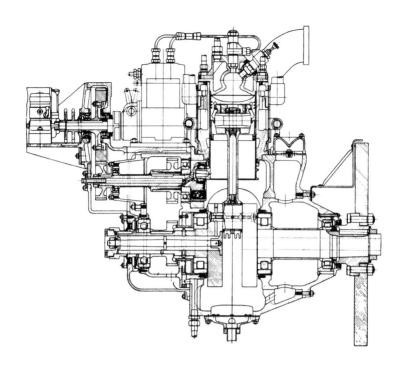

Longitudinal section of the E65 engine
By permission of Ricardo Consulting Engineers

*The RR-D diesel car speed record engine with cover removed
to show sleeve-valve drive*
By permission of Ricardo Consulting Engineers

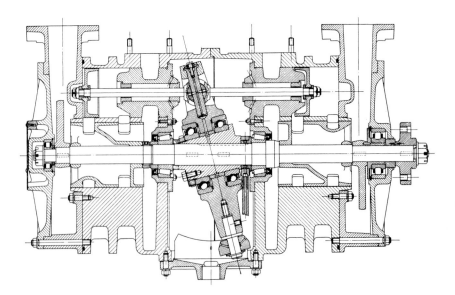

Wellworthy-Ricardo WS 5 compressor
By permission of Ricardo Consulting Engineers

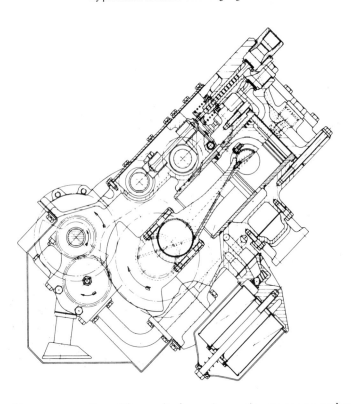

Transverse section of four-cylinder reciprocating steam expander
By permission of Ricardo Consulting Engineers

A Postscript
By Cecil French

In 1990, Ricardo International Plc, the successor to the Company which Sir Harry founded in 1915, celebrated its 75th Anniversary. During its life, the Company has been in the forefront of the developments which have revolutionized the reciprocating internal combustion engine, both spark ignition and diesel.

In writing *The Ricardo Story,* Sir Harry set out to write his autobiography. While it contains some technical detail concerning the Tank Engine and the development of the Turbulent Head, there is little concerning either his or the Company's activities post 1930, since he thought that these had already been more than adequately covered by his published works and those of his collaborators.

In this postscript, I have tried to give some flavour of the wide range of technical activities and achievements, of what many of us still think of as "R & Co," both up to Sir Harry's death in 1974 and subsequently. A full description of the work carried out would require several volumes but fuller details are available to the reader by reference to the many technical papers prepared by Sir Harry and by his colleagues over the years.

Sleeve-Valve Engines
In an earlier chapter, Sir Harry mentions the intensive development of the four-cycle, sleeve-valve petrol engine which was carried out at Shoreham from 1921-1929. This formed the basis of the very successful ranges of Bristol and Napier sleeve-valve aircraft engines, and engines of this type were also built by Rolls-Royce.

In 1936, work started on the development of a very highly rated, two-cycle, gasoline-injected, sleeve-valve petrol engine for aircraft propulsion and the E65 and E54 single-cylinder research units were designed and built. These engines were separately blown and were operated both at Shoreham and at Oxford after the Company had been evacuated from Shoreham in 1940. Experimental running was carried out at very high ratings with a type test being successfully completed at up to 225 lb/in² BMEP at 3000 rpm on the E65 unit.

Toward the end of the programme, short-duration tests were successfully carried out at 354 BMEP at 4000 rpm and it appeared likely that still higher powers would have been achievable, had fuel injection pumps of higher capacity been available.

This work, together with parallel studies being carried out at Derby, formed the basis for the design of the Rolls-Royce Crecy 12-cylinder engine which would have succeeded the Merlin and Griffon had it not been overtaken by the development of the aircraft gas turbine. The Crecy engine, in its developed form, would have been able to produce some 5000 bhp for a gross weight of about 2000 lb with little more frontal area than that of the Merlin or Griffon.

While these aircraft engines gave the ultimate in output, a very wide range of sleeve-valve engines were designed and developed by Ricardo over a period of thirty years or so from 1921-1951. Including two-cycle and diesel versions, some 78 different single-cylinder engine designs were, in fact, carried out.

Multi-cylinder, four-cycle diesel and spark ignition, sleeve-valve engines were designed and were put into production by a wide range of engine builders. An interesting project was a diesel sleeve-valve conversion of a Rolls-Royce Kestrel engine which was ultimately fitted into a car by Captain George Eyston and which set a diesel car speed record of 159 mph which stood from 1936 to 1953.

The Comet Combustion System
From about 1928 onward, the Company became increasingly involved with the application of high-speed diesel engines to road vehicles. Following an early decision to concentrate, at that time, on indirect injection, the 'Comet' swirl chamber system was developed for use in London buses built by the Associated Equipment Company and patents were taken out in 1931.

A range of Comet systems was developed over the years and patents on these variants remained in force in the last country until 1980.

By 1936, 32 companies, in countries from the USA to Australia, had taken out licenses from Ricardo. While the great majority of pre-war Comet engines were employed on trucks and buses, industrial uses, patrol boats, etc., Citroën in France produced a 1.7-litre, four-cylinder engine for use in cars, taxis and small commercial vehicles.

With time, increasing fuel prices and the development of improved fuel injection equipment, direct injection has, of course, taken over from indirect injection for the majority of diesel applications, with the exception of the high-speed, light-duty engines for use in passenger cars and light trucks. While direct injection has now started a small penetration in this area, the vast majority of such engines still employ indirect injection. With the exception of Daimler-Benz, almost all of these engines still employ the Comet Combustion System, 60 years after the filing of the initial patent—a quite remarkable record.

Compound Engines

During the 1930s and 1940s, work was carried out on compound engine concepts, combining a diesel piston engine as a gas generator with a turbine expander. Early work had shown that detonation prevented the use of spark ignition and a comparison of two- and four-cycle diesels demonstrated that two-cycle was superior in terms of power output and specific weight.

Two alternative simple ported single-cylinder engines were constructed, together with a sleeve-valve unit. A very substantial amount of running was carried out and this work formed the basis of the Napier Nomad prototype aircraft engines which were ultimately built, developed and flight tested with results that agreed closely with Ricardo's predictions.

As with the gasoline-injected, two-cycle engine, however, gas turbine developments prevented any production use of compound engines and this work ceased in 1952. Since that date, however, we have come back to it at least twice, once in the original spark ignition form for passenger cars and once for high-power, long-range helicopter applications.

Gas Turbines

While the Company's main involvement has been in the area of piston engines, it played a significant part in the development of the fuel systems for the prototype Whittle gas turbine engine. Fuel burners were manufactured by the Company and then overspeed governors and barometric altitude fuel control devices were both designed and developed and manufactured in prototype quantities. While no further work was carried out on large gas turbines after 1946, in the early 1950s a heavy fuel burning combustor was developed in association with the Shell Company.

A further project concerned the design of small radial inward flow and axial turbines. A relatively large number of design variants were produced in collaboration with the National Gas Turbine Establishment at Pyestock and were tested on a high-speed dynamometer at speeds of up to 80,000 rpm. These tests were ultimately extended to include pulsating flow to simulate turbocharger operation.

The last gas turbine tests to be carried out by the Company involved the testing of ceramic turbine rotors in a gas turbine which had been built for laboratory teaching purposes, employing turbocharger components and designed and developed at Shoreham for the Company's manufacturing subsidiary, G. Cussons Ltd.

Submarine and Torpedo Propulsion

While some of the Company's early work had been directed toward engines employed in submarines and in landing craft and patrol boats, it was not until 1948 or so that any substantial development was carried out directly for the British Admiralty. A test shop with test cells was built at Shoreham by the British Government and operated by Ricardo.

A number of projects were carried out, some of which used High Test Peroxide (HTP) as an oxidant. For submarine propulsion, it was proposed to use Recycle Diesel Engines. A number of experimental power plants were constructed and the final test involved the "dunking" into Shoreham Harbour of a complete power plant fitted into a pressure vessel.

Probably the most "exciting" project was aimed at a power plant for a torpedo employing an inverted cycle where the aspirated charge was vaporised diesel fuel and HTP was injected directly into the cylinder at the appropriate time for combustion to commence. Since HTP mixed with a hydrocarbon liquid forms a high explosive, spectacular results ensued if liquid HTP came into contact with lubricating oil in the piston ring grooves!

Compressors

While from an outsider's point of view Ricardo's is thought of as an engine company, a number of special-purpose compressors have been designed and developed over the life of the Company. These included three-stage air compressors for transportable oxygen plants for use in aircraft to provide oxygen for bomber crews in 1942 and a larger mobile oxygen plant for use in welding by the British Army in 1950.

A two-stage oxygen compressor, compressing to 2600 lb/in² with water lubrication of the pumping cylinders was also designed and developed and subsequently built in some numbers by the Hamworthy Engineering Company for use in aircraft carriers.

Piston-type wobble plate compressors with double-acting cylinders arranged around and parallel with the driving shaft with a rotary valve were also developed in the early 1950s. While the original design was intended for use as an engine supercharger, the production unit found quite wide use as a source of air for unloading cement powder and flour where the freedom from oil carryover in the air, low noise level of the unit and low pulsatory levels of the air offered particular advantages.

A later design of two-stage air compressor, fitted with piston and/or poppet valves supplied air at 100 lb/in² and ran at up to 3600 rpm. This was intended for direct drive by a high-speed, automotive type of diesel engine but while the unit performed successfully, it did not succeed in competition with oil sealed rotary compressors which came on to the market at about the same time.

Steam Engines

Sir Harry had always been interested in steam engines, as exemplified by his Rugby School Motor Cycle. In 1952, he had the opportunity of designing and developing a small steam power plant of transportable form to be used in the underdeveloped world such as India and to "live off the land," burning coal, straw, dung, brushwood, etc.

A two-cylinder, single acting engine with a simple cast aluminium boiler was designed and developed and demonstrated and two small prototype batches were commissioned by the National Research Development Corporation. While tests on these were successful, with the ready availability of small industrial petrol engines in all parts of the world, together with their fuels, it became clear that there was unlikely to be a market for the steam engine.

Steam engines were returned to early in the 1970s when the United States Environmental Protection Agency sponsored a number of projects aimed at the use of vapour cycle engines in order to reduce vehicle exhaust emissions. Ricardo's were a partner in the project which used water as the fluid, with a reciprocating expander.

The steam engine or expander was Ricardo's responsibility and a four-cylinder unit with poppet inlet valves and ported exhaust was designed and prototypes were procured. In order to achieve variable cut-off, each cylinder employed two inlet valves in series, one to control the start of admission and one to control the end. The two valves were operated by different cam shafts and cut-off could be varied by varying the phasing between these two cam shafts.

While low exhaust emission levels were achieved, the fuel consumption of the vehicle was substantially worse than that of vehicles fitted with conventional engines and the pressure on fuel economy, which resulted from the 1973 Middle East fuel price increases, resulted in the abandonment of all three vapour cycle projects.

Stirling Engines

Hot air engines were widely used toward the end of the last century but had been displaced by the internal combustion engine. During the 1930s, however, the Phillips Company in Holland started to develop high-pressure Stirling engines, employing hydrogen as a working fluid. This work continued after the war with larger power units for cars and buses.

The United Stirling Company of Malmo in Sweden was set up to continue work in this area and Ricardo's carried out the mechanical design of the first United Stirling P40 engine and also manufactured a number of experimental units for United Stirling.

Fundamentals of Combustion

In the early 1940s, after the gap of some 20 years since Sir Harry had carried out his classic studies on detonation and pre-ignition, Ricardo's received a contract from the Shell Company to pursue further fundamental studies into the mechanisms involved in detonation, employing the more advanced techniques which had by then become available.

By using high-speed sampling valves derived from Atlas electrically operated injectors, it was possible to follow the chemical nature of the knocking process. By this means, it was possible to forge a link between the engine phenomena being studied at Shoreham and the more fundamental work on combustion which was being carried out elsewhere on laboratory-type apparatus.

Subsequently, the work was extended to cover pre-ignition. The method developed for this involved the use of artificial means of producing a hot spot which could be heated by external means independently of other engine conditions.

Studies of diesel combustion employing high-speed photography also commenced at Shoreham early in the 1950s. Here, too, this work was later supplemented by the use of high-speed sampling to follow, in this case, the mechanisms involved in the production of exhaust pollutants. Still later, Ricardo's were early users of laser anemometry to study air motion and its relevance in diesel combustion.

266

Company Honours

Sir Harry was created a Knight in 1948. Two of his successors as Chairman of the Company, Jack Pitchford and Diarmuid Downs, were honoured by the award of the C.B.E. (Companions of the British Empire) and Diarmuid Downs was further honoured by a knighthood in 1985 and also by election as a Fellow of the Royal Society in the same year.

The Company has provided four Presidents of the Institution of Mechanical Engineers—Sir Harry himself in 1947, Jack Pitchford in 1960, Diarmuid Downs in 1978, and the present author in 1988—and it has also provided Presidents of both major international engine societies, FISITA for automobile engineering and CIMAC for larger diesel and gas engines and gas turbines—Jack Pitchford and Diarmuid Downs being Presidents of FISITA in 1961-63 and 1978-80, respectively, and the present author of CIMAC in 1983-85.

During his lifetime, Sir Harry was awarded a number of medals, honorary membership of professional institutions and honorary degrees including the James Watt International Medal in 1953. He delivered the Horning Lecture to the American Society of Automotive Engineers in 1955.

He died in 1974 in his 90th year, and until shortly before his death, he came to the laboratories at Shoreham several times a week where he particularly enjoyed discussing technical matters with his younger colleagues, including many who were at the start of their professional careers. The Company which he founded had become a Public Company in 1962 and continued. In 1990, it merged with a somewhat younger company, SAC International, which had been founded in 1961 and which carried out somewhat similar design work but in other fields, primarily aircraft design, gas turbines and robotics. The merged company is now known as Ricardo International, thus perpetuating the name.

Biographical Notes

Chorlton, Alan

Alan Chorlton grew up in the industrial environment of the north of England, broken by one interval in Russia in his early 20s as engineer to a large textile plant. During World War I he was Deputy Controller of Aero-engines to the Ministry of Munitions and it was in this capacity that he first met Harry Ricardo.

Chorlton was a pioneer, advocate and talented designer of the automotive (high-speed) diesel, whose work included railcar motors for North America. Probably his crowning achievement, though tragically short-lived, was the design of the engines for the R101 airship, described by Ricardo as ". . . a remarkable performance, for no one had (at that time) produced a diesel engine of anywhere near comparable power and weight."

In the early 1930s Chorlton entered politics, in which he remained until his retirement at the end of World War II.

Clerk, Sir Dugald

Clerk was a pioneer of the internal combustion engine and an early influence on Harry Ricardo. He was a Scot who had spent his younger days in industry in Glasgow, and had built his first gas engine in 1876, when 22 years old. He was the inventor of the Clerk-cycle two-stroke engine in 1879. He studied chemistry and physics, first in Glasgow and then in Leeds, Yorkshire, where he remained for awhile to lecture under Professor Thorpe. He then returned to work on engine development, and was one of the first to realise the significance to engine combustion of flame propagation and turbulence. From 1902 he was engine designer and director of the National Gas Engine Company, and in 1908 was elected a Fellow of the Royal Society for his scientific and engineering work. During World War I he was seconded the Admiralty as director of engineering research. It was mainly for his work here that he received his knighthood.

Dugald Clerk was keenly interested in the motor car, and served on the technical committee of the Royal Automobile Club. As Harry Ricardo recalls, he was noted for his memorable lectures to the Institutions of Mechanical and of Civil Engineers in London, being an active member of both bodies.

Green, Major F.M.

Though Ricardo is less than complimentary about Green's first efforts at aircraft engine design, subsequent work with S.D. Heron at the Royal Aircraft Factory, Farnborough, produced a practical and reliable Vee engine that was used in quantity in World War I. Green left Farnborough following O'Gorman's departure in 1917 and became chief designer to Armstrong Siddeley Motors, where, with Heron, he continued the work on the 14-cylinder radial engines that he had begun at Farnborough.

He was involved in the formation in 1920 of Sir W.G. Armstrong Whitworth Aircraft Ltd, and with another ex-Farnborough colleague John Lloyd as chief designer, was responsible for the Siskin fighter that served the RAF in the early inter-war years, and for the groundwork on the successful Jaguar engine, one of the first supercharged radial aircraft engines. Green retired in 1933, but returned to Farnborough (by that time called the Royal Aircraft Establishment) at the outbreak of World War II.

Halford, Frank

". . . a very capable young flying officer, a certain Captain Halford . . ." is Ricardo's first recollection of Frank Bernard Halford.

When only 21, Halford was involved in the design of the B.H.P. (Beardmore-Halford-Pullinger) 230 hp in-line water-cooled six, one of the outstanding aircraft engines of WWI.

The young Halford had learned to fly before the war, and became a qualified flying instructor. At the end of hostilities (and this is an episode that Sir Harry unaccountably fails to mention) Halford spent two years in America as Ricardo's agent, and was the first to interest Ford in the Ricardo turbulent head. Halford left Ricardo's employment in 1923.

Apart from his love of flying and his flirtation with motorcycle racing in the early 1920s, Halford's working life was dedicated to aero-engines. He was a leading figure in the design of the Armstrong-Siddeley Puma, and subsequently the Napier engines, including the Rapier, Dagger and Sabre (which was to develop more than 3000 hp). At the lighter end of the scale he evolved the Gipsy range that powered so many de Havilland and other aircraft.

Halford worked closely with de Havilland, sharing premises at Stag Lane, North London. In 1941 Tizard (q.v.) involved him in jet propulsion. Halford was responsible for the H.1 engine (later called the Goblin) that powered the first Lockheed Shooting Stars and Curtiss-Wright XF-15, as well as the first Gloster Meteor, and was the standard engine for the de Havilland Vampire fighter.

When the de Havilland Engine Company was formed in 1944, Halford was appointed Chairman and Managing Director. His Ghost engines powered the first Comet airliners, and shortly before his death he was working on the Gyron axial flow gas turbine of 15,000 lb thrust.

Hopkinson, Professor Bertram
From King's College, London, Hopkinson went to Cambridge University, first to study mathematics, and then to read law. However, on the sudden death of his father, a consulting engineer, he relinquished law to study engineering and so maintain his father's practice. Here he became involved in the electrification of tramways, and on patent law cases.

When Professor Ewing retired from Cambridge, Hopkinson was appointed as his successor to the Chair of Mechanism and Applied Mechanics. The future of the internal combustion engine was his overriding enthusiasm. He contributed much to the systematic testing of engines, including some original and valuable inventions, and was an inspiration to his students, of whom Harry Ricardo was one of many to determine the course of engine design in Britain.

During World War I, with the rank of major, he was involved in technical work for the Army, the Admiralty, and then for the

Air Ministry with research and development responsibilities at the new testing station at Martlesham Heath, with the young Tizard (q.v.) as his second-in-command. He was only 44 when he was killed in an air crash in 1918, officially commended for ". . . the patriotic self-abnegation with which he devoted his great abilities to the Public Service."

Horning, Harry

Harry Horning was an influential American pioneer of the internal combustion engine, and particularly of lubrication and fuel research. He shared with Harry Ricardo an interest in the problem of engine knock, and helped develop the ASME octane measuring engine. He was co-founder of the Waukesha Motor Company in 1906, and of the SAE's Research Department. He was President of the Society in 1925.

For many years Harry Horning carried the responsibility for obtaining and administering Ricardo's patents in America. He died in 1936. In 1938 the Horning Memorial Award was instituted by the SAE.

Lanchester, Sir Frederick

Lanchester, who was born in South London in 1868, was an engineer and mathematician of great breadth of achievement. He was educated at what are now Southampton University and Imperial College, London. He was an early and influential theorist of flight, his two major books, the *Aerodynamics* of 1907, and *Aerodonetics* (aircraft stability) of 1908, were seminal works of aeronautical science. He worked on the British government's Aeronautics committee from 1909.

In the early 1900s he turned his innovative mind to the motor car, formed the Lanchester Motor Company, and introduced many developments which, though technically successful, were not widely adopted by the cost-conscious motor industry. Nevertheless, with his brother (widely known as Mister George) a significant number of Lanchester cars were produced for the upper end of the market until 1931. Lanchester was equally inventive in sound reproduction, with a number of patents, was a competent musician and published books of poetry. Ricardo considered him ". . . the greatest inventor of his day . . ."

O'Gorman, Mervin

A few years before World War I the British government established at Farnborough in Hampshire a facility to provide the armed forces with observation balloons, dirigibles and kites. It was called the Royal Aircraft Factory, and had grown out of the Balloon Factory. O'Gorman was its first Director, presiding over a small staff of engineers and a modest-sized workforce. He was a far-sighted, persuasive, and sometimes outspoken individual who was not one to toe the official line that favored airships to heavier-than-air machines. He was an early pioneer of motoring, but more important, he was a firm believer in research, and built the Royal Aircraft Factory into a centre of aeronautical research, which, as the Royal Aircraft Establishment, and subsequently RAE Farnborough, it remains today. His influence was crucial to the development of Allied fighting aircraft in World War I.

O'Gorman remained a powerful influence for science and research, with a keen eye for engineering talent. He died in 1958, recognised more by his fellow engineers and scientists, than by the officialdom that was often the target of his critical Irish wit, as Harry Ricardo (see p. 145) well recognised.

Pomeroy, Laurence

Laurence Pomeroy worked from 1905 to 1919 on the staff of Vauxhall Motors, the UK motor manufacturer that in 1925 became part of General Motors. He then worked in the USA for the Aluminum Company of America, but returned to become chief engineer of Daimler.

He was a noted speaker and writer on automobile engineering, and an advocate of improved valve-train design and of the long-stroke engine. From 1937 to 1940 he was general manager at Stag Lane, North London, of the de Havilland Aircraft Company.

Pye, Sir David Randall

David Pye was one of Ricardo's Cambridge contemporaries and student of Hopkinson (q.v.). Having served as a scientific officer attached to the Royal Flying Corps in the latter part of World War I, he returned to lecture at Cambridge University, where he met Tizard (q.v.). In 1925 he was appointed deputy director of scientific research at the Air Ministry under Wimperis, whom he succeeded as director in 1937. In this capacity he became involved in the early development of the gas turbine. His two-volume treatise *The Internal Combustion Engine,* published in the early 1930s, became a standard text. After World War II he was active in engineering education policy, and was Provost of University College, London, until 1951.

Royce, Sir Henry

Henry Royce was the engineer of the Rolls-Royce partnership. He was apprenticed to the Great Northern Railway when he was 14, then worked in gun manufacture and electric lighting while studying at Finsbury Technical College, North London. At the age of 21 he set up a firm to manufacture electrical equipment but soon became interested in the prospects for the motor car, and in 1906 established the famous firm with C.S. Rolls. The reputation for quality was established before World War I, when the demand for aircraft engines determined the future of the company, leading from the Eagle to the Merlin, and to the first allied jets. Royce was noted particularly for the high engineering standards he set. To Ricardo he was ". . . the great perfectionist."

Stanton, Sir Thomas

". . . we had the whole-hearted co-operation of Dr Stanton of the National Physical Laboratory . . ." wrote Harry Ricardo (p. 213). Dr Thomas Stanton was in fact the founder in 1901 and first superintendent of the engineering department of the NPL at Teddington, West London. Here he remained until his retirement in 1930, contributing much to the knowledge of friction and lubrication, including his classic text *Friction,* published in 1923. Before his tenure of office at the NPL, Stanton had worked as demonstrator at the Whitworth Engineering Laboratory under Professor Osborne Reynolds, of Reynolds' Number fame. Stanton was knighted for his services to engineering science. He died in 1931.

Tizard, Henry

Henry Tizard, researcher and administrator, rose to become one of the most powerful figures in British research and development in the inter-war years, and through World War II. He came from an engineering family and was son of Thomas Tizard, hydrographer of the Royal Navy. He read science and mathematics at Oxford University, and went on to Berlin to study under Nernst. It was here that he met Lindemann (later Lord Cherwell, Churchill's scientific advisor).

In 1911 he was demonstrator in Oxford University's electrical laboratory under Sir Ernest Rutherford, and was travelling to Australia with Rutherford for a scientific meeting when World War I broke out. He returned, enlisted in the Army, was recognised for his methods of training recruits, and transferred first to the Central Flying School at Upavon, and then to work on bombsight design for the Royal Flying Corps under Hopkinson (q.v.). Hopkinson was sufficiently impressed to recommend Tizard for work at the new Martlesham Heath experimental field, and then to join him at the headquarters of the Ministry of Munitions. It was while he was at Martlesham Heath that Tizard met Ricardo. When Hopkinson met his death in 1918, Tizard took over Hopkinson's work.

Tizard then returned to Oxford where he worked on problems of fuel compatibility and detonation with Ricardo and Pye, supported by Shell, under Sir Robert Waley Cohen.

With the worsening political situation in Europe in the mid 1930s, the Tizard Committee was set up to provide scientific backing to Britain's air defence. Two of the Tizard projects that were to play a decisive role in the war were RDF (later called radar), and the support for Whittle's work on the jet engine. In 1940, one year into the war in Europe, Tizard organized the joint British/Canadian scientific mission that brought details of British secret research to the USA, including work on the cavity magnetron that made possible centimetric wavelength radar, and less than a year later details of jet propulsion. After the war Tizard remained involved in defence policy, but gave more of his time to the teaching of engineering, and the expansion of engineering facilities at universities in Britain and the Commonwealth. He died in 1959.

Wallis, Sir Barnes

Harry Ricardo encountered Barnes Wallis in the 1920s when Wallis was working on the airship developments so vividly related by novelist Nevil Shute in his autobiography *Slide Rule* (Heinemann). ". . . we came to look upon him as a tower of commonsense, sound judgment and ingenuity" is how Ricardo recalls Barnes Wallis.

Barnes Wallis was trained as a marine engineer at Cowes, Isle of Wight, but turned to aviation soon after joining Vickers, remaining with that company for most of his life. He was particularly noted for his work on aircraft structures, and for the development of the geodetic, lattice-like framework that culminated in the structure of the Wellington bomber of World War II. He also turned his attention to bomb design, including the 10 ton Grand Slam and the bouncing bomb that was used successfully against German dams.

After the war Barnes Wallis became a leading advocate of supersonic flight and variable geometry wings, and remained in the forefront of aeronautical thought until he retired in 1971.

Wilson, Major W.G.

Major Walter G. Wilson was first and foremost a transmissions engineer who is best remembered for the Wilson Gearbox, the semi-automatic epicyclic transmission and forerunner of the automatic transmission that was used for many years and continues in use on London buses and other transport applications.

Wilson's first successful involvement in motor vehicle transmission goes back to the Wilson and Pilcher car of the early 1900s, built after Pilcher had been killed in a gliding accident in 1899, and plans for an engine for powered flight had been dropped. During the World War I Wilson was engaged as chief of design for the mechanical warfare department of the War

Office. He was in charge of the very first tank trials. It was while involved in the design of the transmission system for the first tanks that he first met Harry Ricardo.

He was an active member of the Institutions of Automobile Engineers, Mechanical Engineers, and Civil Engineers. He died in 1957.

Index

Aachen, 147
Ackroyd Stuart Co., 57, 58
Adrian, Lord, 81
Adur, River, 99, 248–9
Aero-engines. *See* Engines
Aeroplanes, 140–6
Aircraft, party politics and, 220
Airship Guarantee Co., 220–5
Airships, 143, 157, 219–25
Alcock, 207–8
Amberley, 249
American White steam car, 139
Andersen, Hans, 14
Andros, Mr., 119
Angus and the Dolphin engine, 103–8
Anti-knock. *See* Knock
Antoinette engine, 141
Anzani engine, 141
Argyll car, 67
Argyll Motor Co., 144
Armistice, 164, 245
Armstrong Co., 33–4, 162–3
Armstrong, Lord, 33, 34
Arrol-Johnson Co., 78, 98
Art Workers Guild, 30
Arundel, 248
Australia, 29
Austro-Daimler engine, 144, 160

Bagnall-White, General, 160–1
Balloons, 143, 157
Barnes, Bishop of Birmingham, 85
Barrington, Mr., 151–3, 155, 158
Barton, Mr., 213
Beale, Mr., 190
Beardmore Co., 160–3
Bellis & Morcom engine, 50
Benz Co., 9–10, 60
Benz engine, 142
Benzole, 127, 188
B.H.P. engine, 160–1, 163
Bickerton, Mr., 116
Bickerton & Day, 172
Blackie, John, 128
Blackie & Sons, 128–9, 147, 241
Bleriot, 141, 142–3
B.M.E. engine, 142

Boats, 247–50
Bombay-Baroda Railway, 124
Borneo, 188, 204, 205
Boscombe, 33
Bourdon tubes, 134
Brabazon of Tara, Lord, 141, 142, 144
Bridge Builders, The, 119
Bridges, 118–19, 121 ff.
Briggs, Commander Wilfred, 151–9, 160, 167, 185
Brighton, 16, 34, 38; Locomotive Works, 108
Brinton, Mary (Baroness Stocks), 26
Brisbane, 29
Bristol Aeroplane Co., 215, 228, 229
Bristow, Cooke & Carpmael, 234–9
British Daimler Co., 65
Britannia Engineering Co., 111–12, 126
Brixham, 249
Brody, Mr., 229
Brooklands, 138; competitions at, 209–12
Brooman-White, Mr., 190
Browett engine, 51
Browett-Lindley, Messrs., 116, 172
Brown, Mr., 200
Bryant, Commander, 154–9, 175–6, 183, 211
Burne-Jones, Sir Edward, 39, 44
Burt & McCollum, 226–7
Busk, Mr., 143, 145
Butler, Cavendish, 98, 103, 151
Butterflies, 253–5

Caldecott Community, 25
Cambridge, 10, 33, 50, 81–95, 114–17, 125, 145, 188, 196, 216, 243; Automobile Club, 82, 85; motor-cycle competition at, 85–90
Cambridge Scientific Instrument Co., 81
Canterbury, 33
Carter, Major, 213
Cave-Browne-Cave, Commander, 167, 182
Chater Lea Co., 82
Cherwell, Lord, 196

'Chitty-Bang-Bang', 209
Chorlton, Sir Alan, 116–17, 183, 220, 224–5
Churchill, Sir Winston, 36; and tanks, 167
Clarkson, Messrs., 140
Clerget engine, 140
Clerk, Sir Dugald, 68, 78–9, 92, 96, 117–18, 128, 146, 183–5, 196
Cobra, H.M.S., 130–1
Cockerill, Jock, 56
Cockerill-Ougree, Messrs., 56
Cody, Colonel, 143, 144
Cohen, Sir Robert Waley, 187–8, 190, 204–5, 208
Colchester, 99
Combermere, Lord, 190
Commercial vehicles, 139–40
Condor engine, 199, 220, 223
Cooke, George, 234–9
Crankshafts, 215–17
Cripps, Sir Stafford, 234–9
Crossley Brothers, 91, 115, 116, 133, 171, 172, 176, 229
Crossley gas engine, 46, 54
Cycle-cars, 109–11
Cylinder wear, 217–19

Daimler Co., 9–10, 60, 145; engine, 170, 177–8
Dalby, Professor, 118
Dartmouth, 250
Darwin, Erasmus, 151
Darwin, Sir Charles, 196
Davidson, Mr., 116
De Dion engines, 67–8
De Dion Bouton car, 67
D'Eyncourt, Sir Eustace Tennyson, 167, 171, 188–93
De Havilland aeroplanes, 164
De Havilland, Geoffrey, 142, 143, 145
De la Rue, 188–93
de Morgan, William, 31–2
Derby, 195
Deutz, Otto, 116
Devonport Dockyard, 106
Discol, R., 212
Distillers Co., 209–12
Dobbie, Mr., 115
Dollis Hill, 182
Dolphin engine, 96–111, 117–18, 126–8, 137, 152–3, 187, 191, 200–2
Don, Alan, Dean of Westminster, 50
Dorking, 25
Doughty, Mr., 153, 159–60
Dreadnought, H.M.S., 130
Duff, Mr., 218–19
Dumbledon, 28

Dumfries, 78, 161–3
Dykes, Mr., 82, 85, 90, 97, 98

Eagle engine, 199, 207
East India Railway, 124
Edward VII, 30
Electricity, 52, 58–9
Ellis, General, 180–1
Ellis, Major, 174
Ellor, Mr., 194
Elswick Naval Shipyard, 33–4
Ely, 95
Engine Patents Ltd., 191, 197, 213–29, 232–42; *v.* Humber-Hillman Co., 233–9
Engines, 51–9; gas, 46, 53–6; steam, 51, 52–3, 58–9, 123; marine, 51–2, 104–6, 136; diesel, 53, 57, 58, 116, 132–6, 177–8, 233–9; oil, 56–8; petrol, 58, 77, 96–111, 136–46, 198–212, 232–3; hot-air, 58–9; internal combustion, 66; pumping, 77–9, 96, 111; aero-, 126–8, 140–6, 152–64, 193–4, 198–212, 224, 225, 229; steam turbine, 130–1; submarine, 134–5; piston steam, 135; hot-bulb, 135–6; long-stroke, 137–8; motorcar, 144, 200; torpedo, 153–9; tank, 170–81, 188–93, 231, 245; testing, 215–19, 226–8; hydrogen-kerosene, 223, 225
English Mechanic, 70
E.N.V. Co., 144
E30/1 engine, 227–8
Ethyl alcohol, 208–10
Ethylene, 221
Evans, Aubrey (Major), 196–7, 200, 231
Everett, Mr., 162–3
Ewing, Professor, 84

Farnborough, 143, 144–5, 151, 164, 182, 193–4, 196, 199, 213, 226
Farren, Sir William, 196
Farrer, Campbell, Messrs., 191–2
Farrer, Porter & Co., 191
Farwell, Mr. Justice, 235–9
Fedden, Mr., 194–5, 215–17
Ferguson, A., 172, 175
Ferranti, Messrs., 185; alternators, 46
Ferreday, Mr., 119, 120
Fiat car, 138
First World War, 10–11, 50, 67, 81, 82, 112, 125, 131, 133, 139, 141, 144, 145, 147–203, 219, 224, 225, 228, 231, 245
Fishing, 247–50
Flying-boats, 153–9, 165, 185–6

Foster of Lincoln, Messrs., 167
Fowey, 250
Fuel economy, 226–8
Fulham, factory at, 31

Gardening, 246, 255
Gardner & Sons, L., 172, 175, 179;
　hot-air engine, 58
Gas Engine, The, 128
Gas, Petrol and Oil Engine, The, 128
General Motors, 238
Germany, 141–2
Gibson, Professor, 194
Gipsy engine, 229
Gladstone, W. E., 34, 35
Gnome engine, 140
Goodenough, Mr., 185, 188–93, 238
Gordon, Hamilton, 97, 100
Graffham, 27, 77, 80, 111, 126, 243
Green, Mr., 194; engine, 144–5
Grimm brothers, 114

Haggard, Rider, 21
Hale, Beatrice, marriage of, 125
Halford, Captain, 160–4, 189, 229; at
　Brooklands, 210–12
Halsey, Miss, 28
Hancock & Galsworthy Gurney, 59–60
Handley Page, 142
Hatchlands, 35
Hayward Tyler, Messrs., 58, 77
Heald, Sir Lionel, 234–9
Henn, T. R., 9
Henty, G. A., 21
Hersey, H., 256
Hesselmann, Dr., 135–6, 147
Hetherington, Harry, 82, 114, 117, 151,
　160, 182, 183, 189, 196, 243; and the
　Dolphin engine, 99–111
Hillman 'Wizard' engine, 233
Hindenburg, F.-M., 149
Hispano Suiza Co., 138
Hives, Lord, 194–5
Hobson, Captain William, R.N., 35–6
Hobson, Eliza (Lady Rendel), 35–6
Hobson, Mrs., 35
Hobson, Polly, 35–6
Holt, G., 172, 175
Hopkinson, Professor Bertram, 10, 80,
　82–95, 97, 98, 115–18, 125, 128,
　145–6, 153, 182, 187, 193, 196–7,
　200, 205, 236; optical indicator of,
　80, 91–5; death of, 197
Horning, Harry, 232, 238
Horley, Messrs., 25; car, 67
Hornsby, Messrs., 116, 172
Hornsby-Ackroyd oil engine, 57–8, 116,
　135–6

Humber-Hillman case, 233–9
Hunsaker, Mlle, 23–4
Hydraulic power, 121–2
Hydrogen, 220–1

Illustrated London News, 42
India, 118–24
Industrial Revolution, 32
Inglis, Mr., 82, 85, 86, 90
Institute of Civil Engineers, 33, 115
Ireland, 28, 30
Italy, 34

Japan, 34
Junkers, Professor, 147
Junkers engine, 147
Jupiter engine, 216–17
Just So Stories, 45

Kelvin, Messrs., 105; engine, 105
Kewley, Mr., 187–8, 205, 207–8
Kidner, Percy, 138–9, 183, 238–40
Kiel, 147
Kinghorn valves, 106–7, 110
Kipling, Rudyard, 44–5, 119
Kipling, Miss, 44–5
Kitchener, Lord, 30
Knight double-sleeve engine, 138, 226
Knock, 93, 126–8, 187
Körting, Messrs., 56, 116; engine, 142
Krupps, 147, 148

Lanchester, F. W., 66, 117, 198
Lancing, 245 ff., 251
Laurence Scott & Co., 111
Laystall Engineering Co., 69, 82
Lear, Edward, 14
Le Mesurier, Mr., 162–3
Le Rhône engine, 140
Lindley engine, 51
Lion engine, 199
Littlehampton, 248
Liverpool Casting Co., 75–6
Lloyd, Mr., 100, 109–11
Lloyd & Plaister, 100–2, 109–11, 119,
　126, 191
Lloyd George, 1st Earl, and the tanks,
　167–8, 174–5, 176
Locomobile, 49–50
Locomotives, steam, 52, 131–2
London, 33, 34, 40, 117–21; in the
　early 90's, 13–25; street lighting in,
　52
London General Omnibus Co., 234,
　235–6
Longemarre carburettor, 87
Lubricants, 213–15

Lucas, Keith, 81–2, 151, 196
Lucy, 25–6
Lusitania, S.S., 130
Luton, 139

McKechnie, Mr., 133–4, 200
Mahle, Messrs., 231
M.A.N. Co., 53, 116
Marconi, Guglielmo, 184, 185
Martlesham Heath, 193, 194–5, 196, 205
Mary (nurse), 13–14
Mason, George, 38–45; the Misses, 38–45; Thomas, 38–45
Mather & Platt, 116
Mauretania, S.S., 130
Mechanical engineering at turn of century, 51–68
Methane, 221
Methyl alcohol, 208–9
Mercedes Co., 158; engine, 142
Metallic Valve Co., 106
Metropolitan Carriage and Wagon Works, 176
Metz, 149
Michel thrust block, 51, 131
Midgley, Mr., 239
Miesse car, 139
Minerva engine, 66
Mirrlees, Mr., 116
Mirrlees, Bickerton & Day, 116, 133, 171–2, 176, 179, 229
Model Engineer, 71
Mons, retreat from, 149, 150
Morris, William, 31
Mort, Mr., 143–4
Motor-cars, 48–50, 60–8, 137–40, 245; Benz, 61–5; Panhard, 65; De Dion, 87; British Daimler, 101; Lanchester, 101; Napier, 101; Wolseley, 101; Dolphin, 192; racing, 209–12
Motor-boats, 247–50
Motor-cycles, 82–4, 136–7; racing, 210–12
Motor Manufacturing Co., 67
Motor-tricycles, 66
Moult, Mr., 229
Mudfish, The, 248
Murray, Workman & Co., 119–20, 124

Napier Co., 138, 228, 229
Naples, 34
Napoleonic Wars, 28
National Gas Engine Co., 116, 172
Nernst, Herr, 205
New Engine Co., 143–4
New Zealand, 35
Newmarket, 86

Nicholson, Mr., 111
Norman, Mr., 164, 194
Norwich, 111

O'Gorman, Mr., 143, 144–5, 194, 196, 199
Okehampton, 32
Omnibuses, 140
Onions, Mr., 116
Optical indicators, 80, 91–5
Orde, Edwin, 162–3
Orfordness, 196
Ormandy, Dr., 127, 187
Otto Co., 53
Oxford, 32, 34, 205, 278

Palmer, Sir Frederick, 124–5, 150, 182–3, 192–3
Panhard Co., 60
Parsons Co., 105
Parsons, Sir Charles, 118, 184, 185, 216
Passenger, Mr., 31–2
Patents, 231–42
Pearl, 249–50
'Penstone', 245–52, 255
Peter Brotherhood, Messrs., 153–9, 161, 172, 175–6, 229
Peterborough, 153–9, 175–6
Petrol, 186–8; tests on, 187–8, 204–8; for racing cars, 209–12
Peugot Co., 138
Pike, Mr., 200
Plaister, Mr., 100
Plunkett, Mr., 161–2
Plymouth, 32, 36, 250
Pneumatic tyres, 59
Poison gas, 185
Pomeroy, Mr., 117, 137–9, 176, 198, 241
Poole, 250
Portland, 32
Portsmouth, 33
Power-stations, 132
Pozzuoli, 34
Pratt & Whitney, 194
Price's Heavy Gas Engine Oil, 88
Priestman oil engine, 57–8, 222
Pullinger, Mr., 161
Puma engine, 163
Pye, Sir David, 196, 205, 208, 213, 246–7

R.A.F. engine, 144–5, 152, 194, 198
Rateau, Professor, 199
Reigate, 63; Hill, 63–4, 65
Renault engine, 141
Rendel, Arthur, 36, 102

Rendel, Catherine Jane (Mrs. Ricardo), 26, 32
Rendel, Connie, 26, 36
Rendel, Edith, 25, 125
Rendel family, 25–6, 32–7, 113–14
Rendel, George, 33–4, 36–7, 51
Rendel, Hamilton, 33, 34, 36–7
Rendel, Harry, 36, 113–14; death of, 114, 115
Rendel, Herbert, 36, 101, 103, 107
Rendel, James Meadows, 32–3
Rendel, Leila, 25, 125
Rendel, Lewis, 33
Rendel, Lord, 33, 34–5, 60, 65
Rendel, Robin, 26
Rendel, Sir Alexander, 25–8, 33, 35, 60; 113–14; death of, 182
Rendel, William, 36; death of, 113
Rendel & Robertson, 113–24
Rendel, Palmer & Tritton, 124, 193
R.H.A. engine, 163–4, 196
Ricardo & Co., 191
Ricardo, Angela, 246
Ricardo, Anna, 17, 18, 23–4, 243, 244–5, 250
Ricardo, Arthur, 28, 29
Ricardo, Camilla, 250
Ricardo, David, 28
Ricardo, Esther, 23, 244, 250
Ricardo family, 28–32
Ricardo, Halsey, 13–37, 243; and the Dolphin engine, 99; death of, 230
Ricardo, Harry, 28, 29–30
Ricardo, Kate, 125, 147, 244
Ricardo, Lady Beatrice, 125, 245 ff.
Ricardo, Mary, 28, 29
Ricardo, Mrs., 26, 32, 243
Ricardo, Percy, 28, 29, 72
Ricardo, Ralph, 29, 72–3, 75, 77–8, 152, 192; and the Dolphin engine, 98–108; in India, 108
Ricardo, Sir Harry, passim; early life, 13–27; family background, 13–37; education, 14, 23–4, 38–50, 81–95, 114–17; at Rottingdean School, 38–45; at Rugby School, 45–50; early endeavours of, 69–80; pumping engine of, 77–9, 96, 111; at Cambridge, 81–95, 114–17; makes his first motor-cycle, 82–4; and the Dolphin engine, 96–111; starts his career, 113–29; marriage, 125; and his book, 128–9, 159, 240, 241, 251; forms his company, 182–97; elected F.R.S., 230
Rickettswood, 25–7, 32, 35, 63, 64, 67, 77, 78, 243, 251
Rider hot-air engine, 58

R.N.A.S., 151–60, 165, 167, 172
Road transport, 59–68, 131–2, 135–6
Robertson, Mr., 113–14, 115, 120; death of, 124
Robinson, Mr., 80
Roe, A. V., 142
Rolls, Hon. C. S., 60
Rolls-Royce, 59, 151, 228, 229
'R100', 220, 223, 224, 225
'R101', 220, 224, 225; disaster to, 225
Roots blowers, 143–4
Rothenburg, Mr., 82, 85, 86
Rothschild family, 28
Rottingdean School, 38–45, 46, 71
Rowledge, Mr., 176, 194–5, 198
Royal Aircraft Establishment, 143, 144, 213. See also Farnborough
Royal Tank Corps in France, 180–1
Royalties, 231–42
Royce, 59, 66, 195–6
Royston, 86, 89
Rugby School, 29, 31, 39, 45–50, 73–6, 80, 82, 114; Engineering Society, 46, 49–50
Russian Revolution, 185
Russo-Japanese War, 37

San Jose, 248
Sassoon, Hamo, 100
Sassoon, Michael, 95, 97–101, 151
Schneider Trophy Race, 250
Science in Writing, 9
Scotland, 50
Seaplanes, 165
Second World War, 257–8
Sequin, M., 140
Sentinel Co., 140
Serpollet car, 139; flash boiler, 74
Shaw, Lieutenant Francis, 165–6
Shellhaven, 187
Shell Petroleum Co., 11, 187–8, 190, 191, 208, 209–12, 213–15; Racing Spirit, 212
Shoreham-by-Sea, 11, 137, 153, 192, 200, 204, 213–29, 240, 247 ff.; and the Dolphin engine, 99–112
Short, Messrs., 155
Siddeley Co., 163
Siegener Co., 56
Single-sleeve valve, 226–7
Smith & Son, 69
Southampton, 33, 104, 131
Southwell, Sir Richard, 196
Sperm oil, 88, 90
Stanton, Dr., 213–15, 217
Staples, Mr., 140
Steam cars, 49–50, 139–40
Steam coaches, 59–60

Stern, Sir Albert, 167–8, 174
Stirling, Reverend, 58
Stockholm, 134–5, 147–8
Stocks, Baroness, 26; John, 26
Strand Magazine, 42
Stumpf, Professor, 52
Submarines, 225–6
Sulzer Co., 53
Sunbeam Co., 138, 144
Sunderland, 33
Supercharging, 199
Swinton, Campbell, 118, 183–6, 188–93; death of, 238

Tanks, 10–11, 165–81, 182, 183, 231, 245; Mark IV, 165, 177–9, 190; history of, 166–70; first action of, 169–70; Mark V, 173, 175, 176–9; demolition of, 180–1; fuel for, 186
Taylor, H. B., 226
Taylor, Sir Geoffrey, 196
Telford, Thomas, 32
Templar, Mr., 49, 74, 75, 80
Tetra-ethyl-lead, 239
Theatres, 243–4
Thomas Tilling Co., 236
Thornton, Mr., 98–9
Thornycroft Co., 105, 140
Thornycroft, Oliver, 82, 95, 97, 100, 114, 151, 182, 183, 189, 246–7
Thornycroft, Sir Hamo, 30, 97
Thornycroft, Sir John, 130–1
Tilling, Reginald, 236
Tilling-Stevens Co., 229
Times, The, 162
Tizard, Sir Henry, 194–5, 196, 205, 208, 213, 219, 246–8, 251
Tom Brown's Schooldays, 45
Tottington Manor, 252–8; greenhouse at, 256–7
Trawling, 247–8
Treaty of Waiting, 35
Tritton, Sir Seymour, 114, 120, 150, 165, 176–7, 182–3, 192–3
Tritton, Sir William, 167
Triumph motor-cycle, 211–12
Triumph-Ricardo engine, 212
Tsushima, Battle of, 37

Tudsbury, Dr., 115
Turbinia, 184
Turbines, steam, 51–2
Two-Stroke Engine Co., 98–111, 126
Twyford, 29

Uganda, 36
Uniflow engine, 52–3
U.S. Patent Office, 231–2, 235

Vauxhall Motors, 117, 137–9, 229, 238, 241
Vickers, Messrs., 133–4, 220
Victoria, Queen, 35
Victoria Nyanza, Lake, 36
Viper, H.M.S., 130–1
Vox car, 110–11

Waller, Mr., 255 ff.
Wallis, Barnes, 220, 222–3
Walton-on-Thames, 11, 29, 125–6, 152, 183, 200, 231, 236, 244 ff., 254
Watson, Trevor, 235–9
Waukesha Motor Co., 232
Wedgwood, Felix, 102
Wedgwood, Reverend, 28–9
Welsh, Tony, 49, 81, 114, 151; death of, 50, 162
West Wittering, 195–6
Westinghouse gas engine, 54
Weyburn Engineering, 97
White, Grahame, 142
Whitehead, Mr., K.C., 234–9
Whitelaw, Robert, 38, 45–6, 50, 76
Whitten-Brown, 207–8
Willans & Robinson, 49, 59, 74, 80, 91
Willans, Kyrle, 49
Willans central-valve engine, 49, 51, 52, 54, 59
Wilson, Mr., 167, 173, 176–7
Windeler, Mr., 116, 171–2, 183
Wolseley Co., 144
Wolverhampton, tank tests at, 169–70, 172–3
Wright brothers, 85, 140, 152

Zeppelins, 219–20; raids by, 160